T0318689

R AND PYTHON FOR OCEANOGRAPHERS

R AND PYTHON FOR OCEANOGRAPHERS

A Practical Guide with Applications

HAKAN ALYURUK
Institute of Marine Sciences and Technology,
Dokuz Eylul University, Turkey

Elsevier
Radarweg 29, PO Box 211, 1000 AE Amsterdam, Netherlands
The Boulevard, Langford Lane, Kidlington, Oxford OX5 1GB, United Kingdom
50 Hampshire Street, 5th Floor, Cambridge, MA 02139, United States

Library of Congress Cataloging-in-Publication Data
A catalog record for this book is available from the Library of Congress

British Library Cataloguing-in-Publication Data
A catalogue record for this book is available from the British Library

ISBN: 978-0-12-813491-7

For information on all Elsevier publications visit our website at https://www.elsevier.com/books-and-journals

Publisher: Candice Janco
Acquisition Editor: Louisa Munro
Editorial Project Manager: Hilary Carr
Production Project Manager: Prem Kumar Kaliamoorthi
Cover Designer: Miles Hitchen

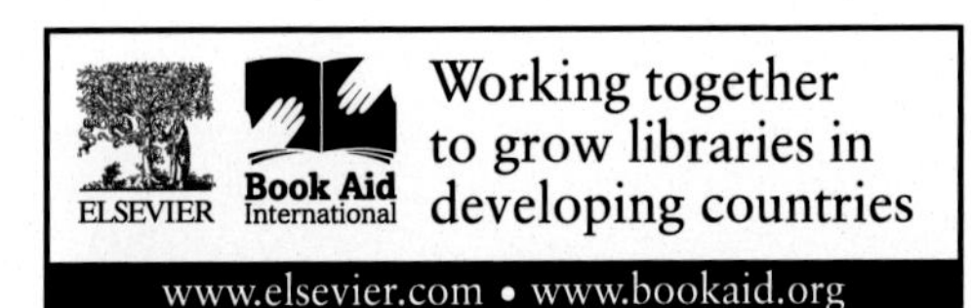

Typeset by SPi Global, India

Contents

CHAPTER 1

Introduction to R and Python

1.1 Introduction to R

R is a language and package-based software environment that provides a wide variety of functions for statistical computing and graphics. R is a project of GNU, and its root depends on S language with some differences. S language and environment was developed by John Chambers and his colleagues at Bell Laboratories (formerly AT&T and now Lucent Technologies). While S language provides programmable research tools for statistical analysis, R has evolved as an Open Source alternative of the S language. R was initially developed by Robert Gentleman and Ross Ihaka at the statistics department of the University of Auckland. Its name "R" comes from the first letters of its developers. Besides its statistical capabilities (such as linear and nonlinear modelling, classical statistical tests, time-series analysis, classification, clustering, etc.), high-resolution publication-quality plots can be produced with full control of the users. R is distributed under the terms of Free Software Foundation's GNU GPL license. Thanks to terms of this license, source codes of both base distribution and external packages with GPL license can be accessed, modified, and re-distributed by the user. Therefore, data analysts can view and contribute to source codes of these packages by accessing the running algorithms and adding their specific functions. As a result, scientists and students at universities can use R freely and without a license fee. The use of free software in science encourages and improves worldwide collaboration on data-analysis approaches. Also, R is used in teaching statistics at undergraduate and graduate levels at universities since students can freely use these statistical tools [1, 2].

The base distribution of R is maintained by the R Development Core Team since 1997. In addition, a huge group of volunteer developers contribute to the project by developing new packages or upgrading existing add-on packages to improve their functionalities. The project's website is the main official information source and can be reached from Fig. 1.1:

https://www.r-project.org

R and Python for Oceanographers
https://doi.org/10.1016/B978-0-12-813491-7.00001-4

Fig. 1.1 A view of official R-Project website.

1.2 R environment

R statistical environment system is composed of two main parts: base-distribution system and add-on packages.

1.2.1 Base system of R

The base system consists of core functions required to run R language. Also, core statistical functions (stats package) and graphical tools (graphics package) are included as dependencies of other add-on packages. Core packages included in the installation media of base system cover standard statistical functions, such as statistical tests, linear models, non-linear models, data-handling functions, clustering functions, and survival-test functions. A list of packages within the base system (included in R version 3.4.0) is given below.

A list of base distribution packages:

base	methods	tools
compiler	parallel	translations
datasets	splines	utils
graphics	stats	
grDevices	stats4	
grid	tcltk	

1.2.2 Add-on packages

Add-on packages are developed and maintained by the user community. These packages contain many advanced statistical, computational, and graphical functions. Some of the packages have dependencies on other packages.

Examples of add-on packages:

agricolae	maps
car	plyr
ggplot2	psych
lattice	

1.3 Installation of R

1.3.1 Installation of R base software

R is an open source software, and it can be downloaded and installed freely. Latest/older versions of R software for Windows, Linux, or Mac OSX operating systems could be downloaded and installed from http://cran.r-project.org/ (Fig. 1.2).

In Linux systems, like Ubuntu, the latest version of R base (r-base) environment can be installed from Software Center by adding official repositories of R-CRAN (Comprehensive R Archive Network) to the repository list. In Debian systems, latest R version can be installed from Package Manager after adding appropriate CRAN repositories. In Mac OSX systems, the latest version of R can be downloaded from its official website and installed according to the instructions.

1.3.2 Installation of add-on packages

You can install add-on packages from the internet into the package library folder under R installation directory or any user-defined library folder as below:

install into default package library folder

```
install.packages("package_name")
```

install into user defined library folder

```
install.packages("package_name", lib='C:\Program Files\R\R-
version\library')
```

Fig. 1.2 A view of R-Project download page.

Or you can install packages from the R console application under the Packages menu by clicking the install packages option. Then, you can select the desired package to install from the drop-down list of add-on packages.

Alternatively, you can install a previously downloaded package from a local file by following these instructions:

1. First download the latest versions of packages with their dependencies from CRAN
2. Create PACKAGE and PACKAGE.gz logs for packages in the directory with these codes:

```
library(tools)

write_PACKAGES("C:/Users/username/r_packages")
```

3. Delete PACKAGE.gz file
4. Run this code to install the package:

```
install.packages("package_name", contriburl="file:
C:/Users/username/r_packages ")
```

Optionally, the latest or test versions of some packages can be installed directly from its Github page maintained by its user:

1. First install devtools package that allows the installation of packages from Github repository as well as it provides many useful tools:

```
install.packages("devtools",
lib="file:C:/Users/username/r_packages")
```

2. Install desired package from its Github repository:

```
devtools::install_github("Github_username/package_name")
```

You can test the installation of add-on packages by loading them from R console with this code:

```
library(package_name)
```

If the code runs without any errors, this means the installation of the package was successful.

1.4 Integrated development environments (IDEs) and editors for R

Commands, functions, and scripts of R language can be edited and saved with lightweight and user-friendly code editors like Notepad++, RKWard, Rgedit, Atom, Vim with Nvim-R plugin, Emacs with ESS plugin as well as with other text editors. Some of these editors provide code highlighting, autocompletion, and syntax error-checking functions. Behind these editors, IDEs like Official R Console and Editor, R Studio, and Tinn-R have additional code execution and plotting capabilities. These IDEs have multi-panel systems within a single application window, allowing the user to manage scripts easier and faster.

1.4.1 Official R console and editor

R base software installation comes with an R editor for writing of commands/scripts and an R console for execution of commands (Fig. 1.3). The console is an interpreter of R language and works like a command line interface such as CMD in Windows and Terminal in Linux systems. It is possible to run saved scripts from a file and to load and save workspace and history of commands. It is also possible to manage packages and to open help files from the main menu bar.

1.4.2 RStudio

RStudio is a free and open source IDE for R language (Fig. 1.4). Compared to official R console, it provides many advantages for editing and running of scripts. RStudio is available for Windows, Linux, and Mac OSX systems. More information about RStudio can be found at its official website:

https://www.rstudio.com/

1.5 Useful R commands

show version number of R and installed OS

```
sessionInfo()
```

show R version

```
getRversion()
```

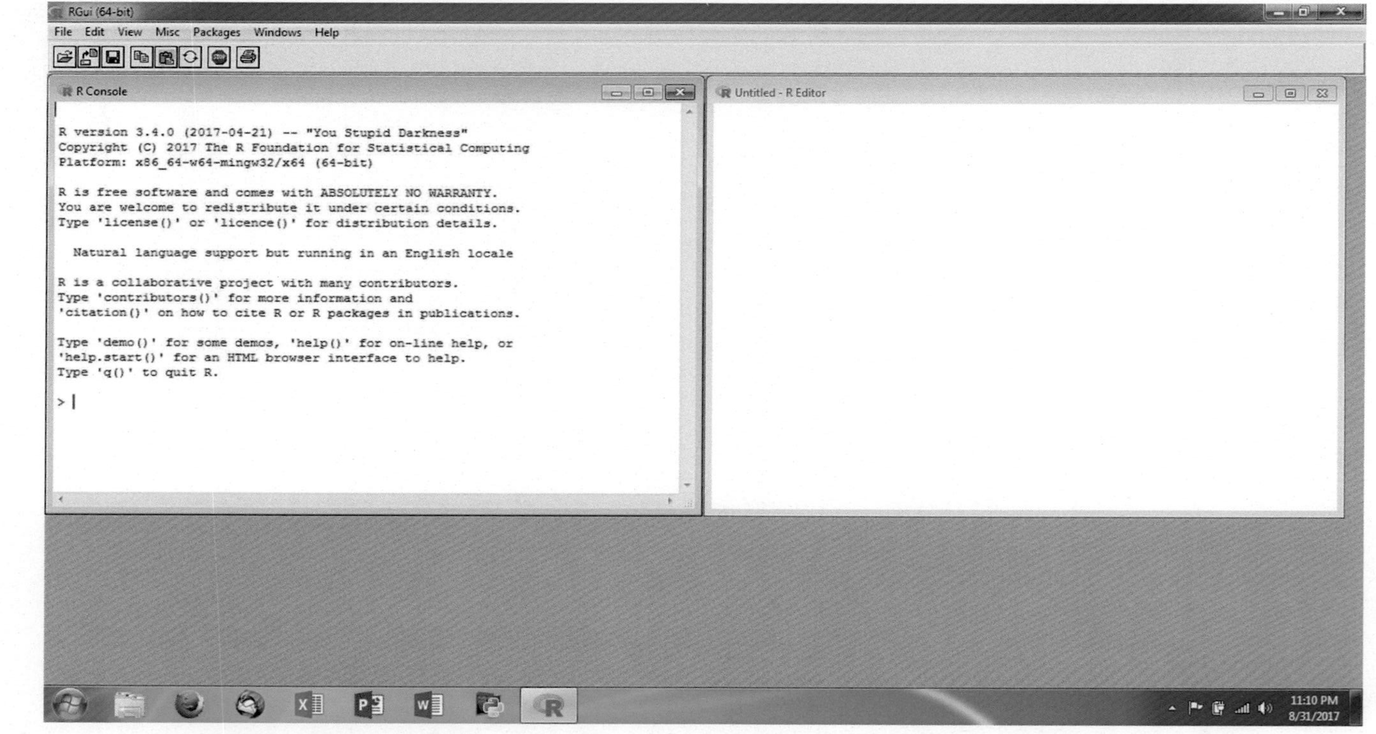

Fig. 1.3 A screenshot of R editor and console.

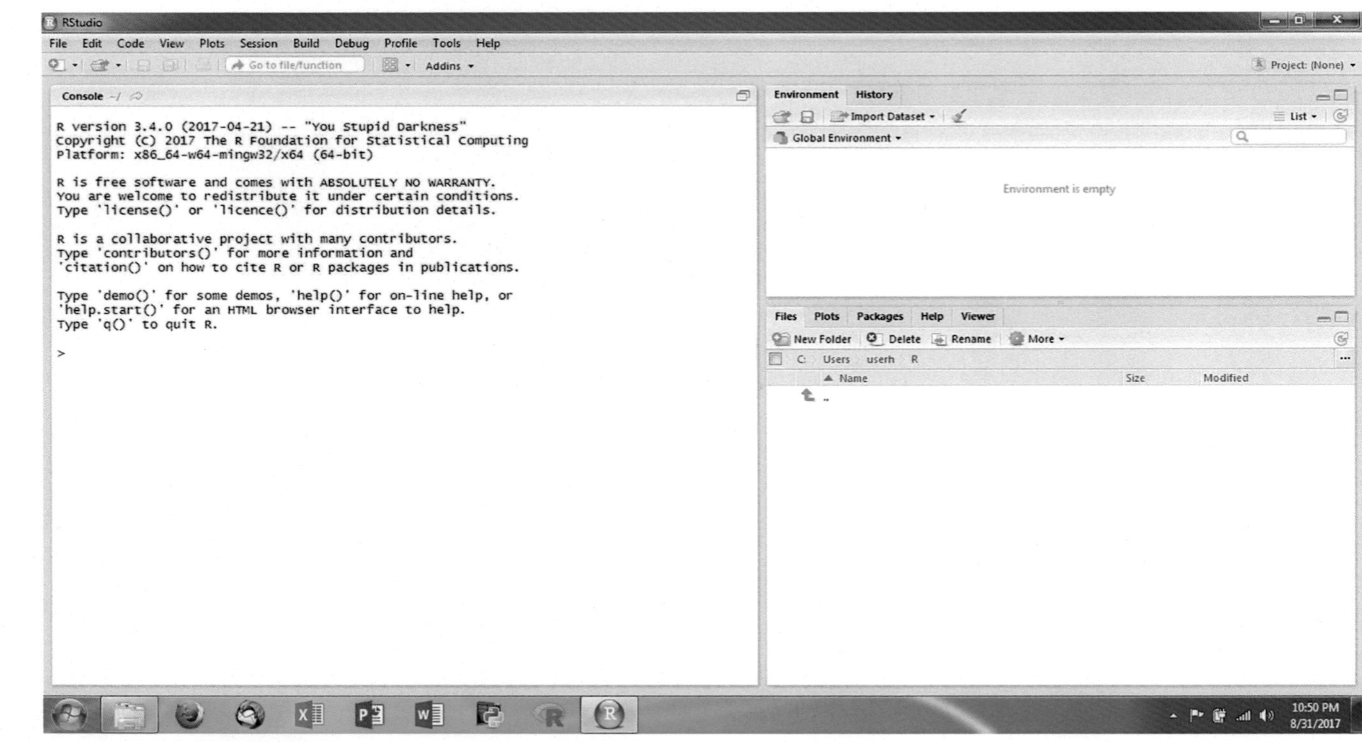

Fig. 1.4 A screenshot of RStudio.

show current working directory

```
getwd()
```

change working directory to a user defined location

```
setwd("C:/...")
```

list all packages in the R package library

```
library()
```

open download mirrors list for package installation

```
available.packages()
```

download R packages from CRAN repository to a local folder

```
download.packages("package_name", "download_directory")
```

update all packages

```
update.packages(checkBuilt=TRUE)
```

show version numbers of R packages, dependencies of packages, recommended packages, license info and builded R version number

```
installed.packages()
```

parses and returns the description file of a package

```
packageDescription("package_name")
```

remove packages

```
remove.packages("package_name", "installed_directory")
```

show license info

```
license()
```

display citation

```
citation()
```

attach packages

```
library(package_name)
```

detach packages

```
detach("package:package_name")
```

list all packages in the default library

```
library(lib.loc = .Library)
```

list loaded packages

```
search()
```

running time of R

```
proc.time()
```

current memory usage

```
memory.size()
```

total available memory

```
memory.limit()
```

terminate an R session

```
q()
```

display warning messages

```
warnings()
```

display the commands history

```
history()
```

load the commands history

```
loadhistory(file = ".Rhistory")
```

save the commands history

```
savehistory(file = ".Rhistory")
```

Save R image

```
sys.save.image("C:\\Users\\username\\...\\filename.RData")
sys.load.image("C:\\Users\\user\\...\\filename.RData")
```

1.6 Getting help for R

Both online and offline help documentations and manuals for R are available. You can start help while using R with help.start() function from the console. For more help, you can use online documentations from the official R website by using these links:

- Help page: https://www.r-project.org/help.html
- Manuals page: https://cran.r-project.org/manuals.html
- Frequently Asked Questions (FAQs) Page: https://cran.r-project.org/faqs.html

1.7 Introduction to Python

Python is an interpreted, general-purpose, object-oriented and high-level programming language. It was conceived by Guido van Rossum in the late 1980s, and its first implementation started in 1989. Python is developed as a successor of ABC language at Centrum Wiskunde & Informatica (CWI) in the Netherlands. In the development of Python, van Rossum also influenced from Modula-3. As a result, Python has been evolved as a new language that includes improvements over ABC and Modula-3 [3, 4].

Currently, the development of Python language is organized and supported by Python Software Foundation (PSF).

Some advantages of Python as a data science tool are:

- It is simple, readable, and easy to learn.
- Python interpreter can run on multiple platforms: Windows, Linux, and Mac OSX.
- It is possible to express functions with fewer codes compared to other languages.
- It is an open source language and can be freely used and distributed.
- There are many libraries for scientific computing.
- It is possible to use Python in combination with other languages such as C or C++.

1.8 Modules and packages in Python

The power of Python language depends on the availability of modules and packages for wide ranges of applications. A module is a file that contains functions and definitions [5]. Modules include pre-defined and special functions, and they are imported at the start of Python scripts when needed. Modules are created to keep functions modular and separate.

A package is a collection of related modules in a single hierarchical structure [6]. Packages can be very complex to include hundreds of individual modules. Some examples of widely used and fundamental scientific Python packages are: NumPy, SciPy, Matplotlib, Pandas, Sympy, IPython, scikit-learn, seaborn, and SciTools.

These packages provide the following functionalities:

- NumPy is a base package for N-dimensional arrays, linear algebra, integration of C/C++, and Fortran codes, etc.
- SciPy is a fundamental library for scientific computing, such as numerical integration and optimization.
- Matplotlib is a 2D plotting library for high quality and publication-ready figures.
- Pandas provides high-performance data structure and data analysis tools.
- Sympy is a symbolic mathematics library.
- IPython is an enhanced interactive console for Python.
- scikit-learn is a machine learning library.
- seaborn is a matplotlib based visualization library for statistical graphics.
- SciTools is a scientific computing library built on top of other widely used packages such as NumPy, SciPy, Gnuplot, etc.

1.9 Python IDEs

There are many IDEs for editing and running python scripts. This multipurpose software provides ease of writing, highlighting and formatting codes as well as finding syntax errors. They also provide autocompletion function and file/path browser for easy handling of script files. Some examples of IDEs for scientific computing are: Python IDLE (Python's Official Integrated Development and Learning Environment), Eclipse, Eric and Spyder. Application previews of Python IDLE and Spyder are given at Figs. 1.5 and 1.6, respectively.

1.10 Installing Python and scientific Python distributions

Python comes pre-installed in Linux and Mac OSX systems. In Windows, Python interpreter could be installed from https://www.python.org/downloads/windows/. In Linux and Mac OSX systems, individual scientific python packages could be added to local python libraries.

In Linux and Mac OSX systems, standalone python packages (installation of SciPy stack given as an example) could be installed by using one of the below options:

1.10.1 Using pip package

To use this option, Python and pip should be already installed on your system.
First, upgrade the pip package:

```
python -m pip install -upgrade pip
```

Then, install the latest version of a module and its dependencies from the Python Packaging Index (PyPI):

```
python -m pip install package_name
```

For example, install SciPy stack using user option:

```
pip install -user numpy scipy matplotlib ipython jupyter
pandas sympy nose
```

Add user installation directory to system's PATH by,

Fig. 1.5 A screenshot of Python IDLE.

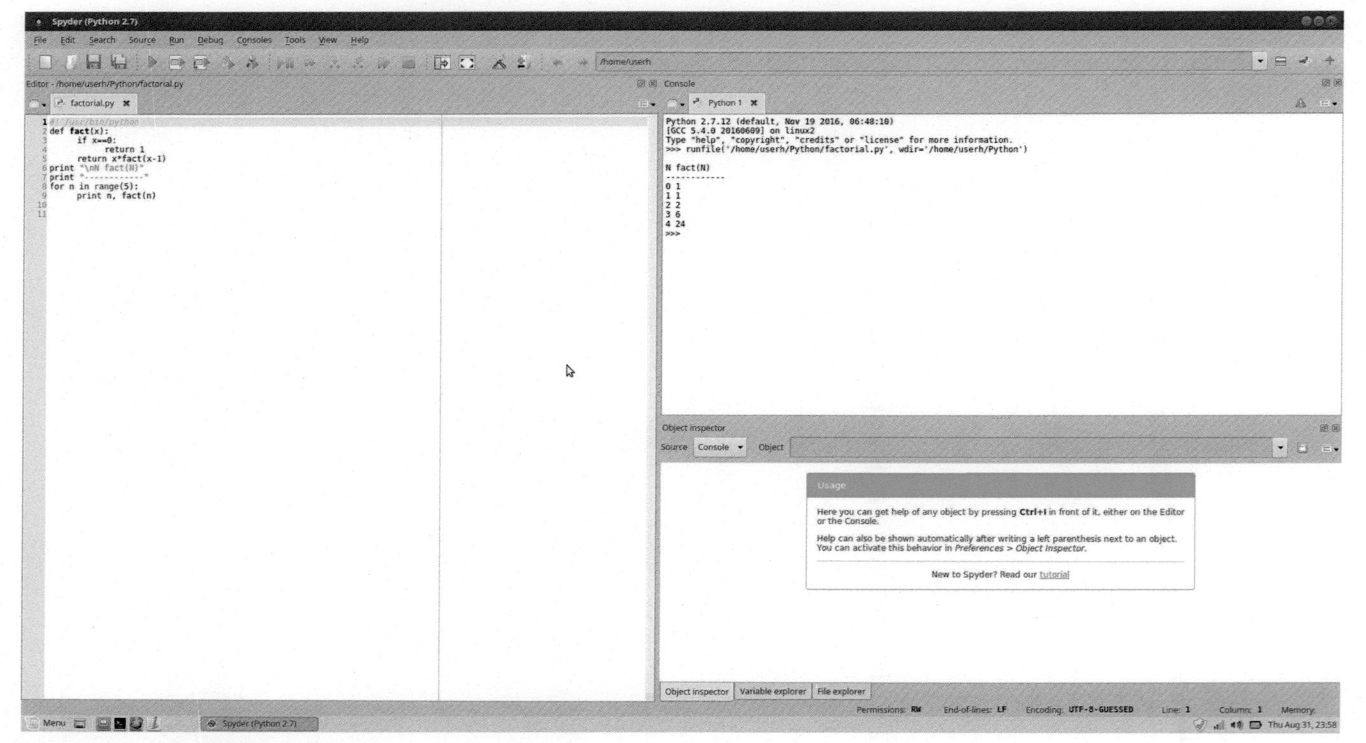

Fig. 1.6 A screenshot of Spyder IDE.

For Linux:
Add below command to the end of your ~/.bashrc file

```
export PATH="$PATH:/home/your_user/.local/bin"
```

For Mac OSX,
Add below command to the end of your ~/.bash_profile file

```
export PATH="$PATH:/Users/your_user/Library/Python/3.5/bin"
```

Change "your_user" with your username and "3.5" with Python version installed on your system.

1.10.2 Using package managers

Depending on the availability of packages and their update status on the repositories of your operating system, this option might install latest/previous versions of packages.

In Linux, you can choose from these options, depending on your distribution:

For Ubuntu or Debian systems,

```
sudo apt-get install package_name
```

For example, to install SciPy stack using this option:

```
sudo apt-get install python-numpy python-scipy python-
matplotlib ipython ipython-notebook python-pandas python-
sympy python-nose
```

For Fedora:

```
sudo dnf install package_name
```

For example, to install SciPy stack using this option,

```
sudo dnf install numpy scipy python-matplotlib ipython
python-pandas sympy python-nose atlas-devel
```

1.10.3 Using source files of packages

If package files are not found at PyPI and at repositories of your operating system, you can use this option. First, download the latest source code of the package to a local directory. Uncompress the installation files from the downloaded source file (usually compressed as tar.gz or zip files). Then, run these codes to install the package:

For Linux systems

```
python setup.py install
```

For Windows

```
setup.py install
```

Detailed instructions for installation of SciPy stack could be found at https://www.scipy.org/install.html. Depending on their availabilities at your system's repositories, other python packages could also be installed by using package managers or by installing from their source files.

For windows systems, scientific python distributions are the easiest solution to use scientific python packages. They provide a great number of packages with a single software installation. Canopy (by Enthought), Anaconda (by Continuum Analytics), and Python(x,y) are among several scientific python distributions that are free to use for educational and academic purposes. Canopy and Anaconda are multi-platform (Windows, Linux, and Mac OS-X versions are available) distributions with the support of the latest package updates, whereas Python(X,Y) is only available for Windows systems with update schedules at longer intervals. An application preview of Canopy Editor is given at Fig. 1.7 as an example.

Some advantages of using these scientific python distributions are as follows:

- Python packages are automatically installed along with Python environment.
- It is possible to manage packages from GUI package managers.
- Python code editor and interpreter are provided in the same application window.

Download links for the latest and older versions of Canopy, Anaconda, Miniconda and Python(x,y). Distributions are given below:

- Canopy: https://store.enthought.com/downloads
- Anaconda: https://www.continuum.io/downloads

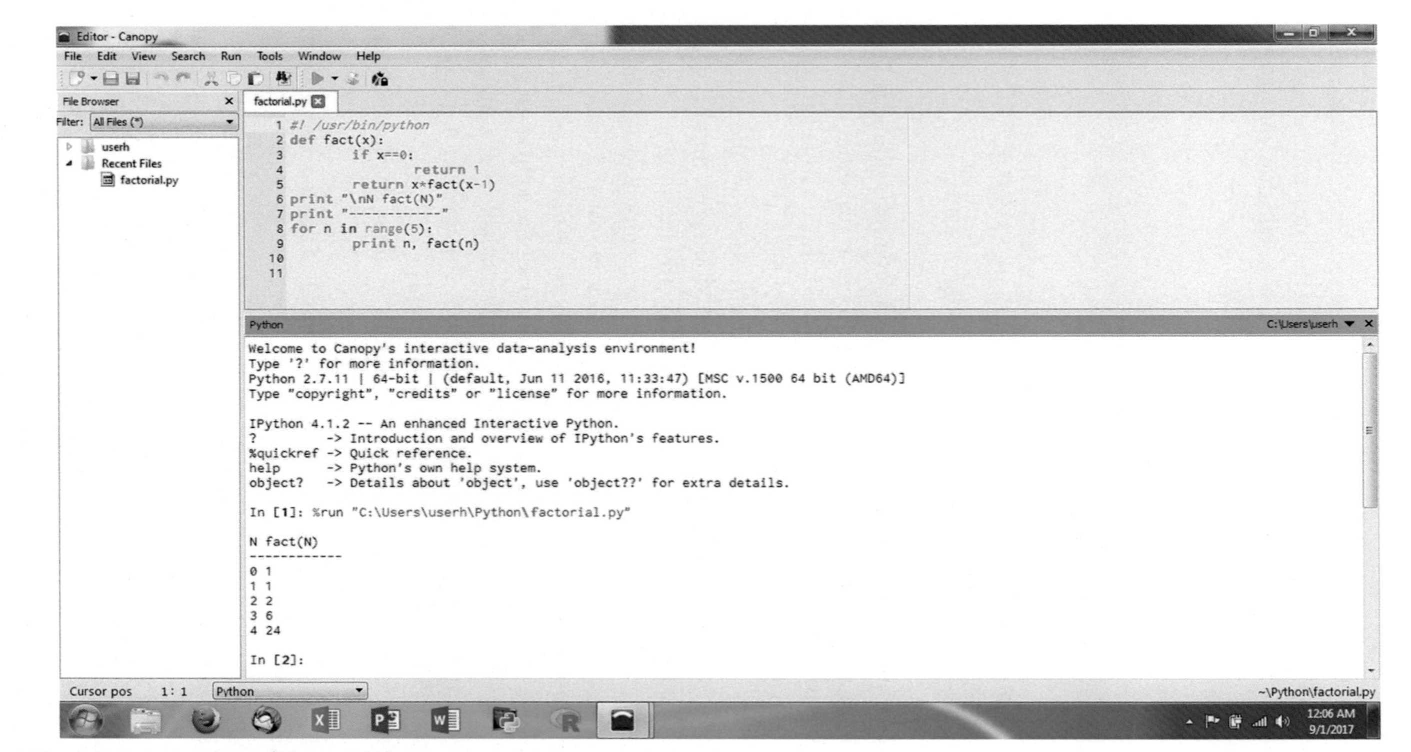

Fig. 1.7 A screenshot of Canopy Editor.

– Miniconda: https://conda.io/miniconda.html
– Python(x,y): https://python-xy.github.io/downloads.html

To test the Python installation:

– Open terminal (in Linux/Mac OSX)/command prompt (in Windows), type python and press enter or click Python icon from the start menu.
– Then, run the following code at the Python shell:

```
import sys
print sys.version_info
```

– If you get the below output, it means that Python (version 2) works without any problem on your system.

```
sys.version_info(major=2, minor=..., micro=..., ...)
```

1.11 Getting help for Python

Python has online and offline help documentations. You can start help while using Python with help() function from the console. For more help, you can use online documentations from the official Python website by using the following links.

- Documentation page: https://www.python.org/doc/
- Help page: https://www.python.org/about/help/
- FAQs page: https://docs.python.org/2.7/faq/ or https://docs.python.org/3/faq/

1.12 Some useful packages and libraries in R and Python for oceanography

In R and Python, some packages useful for oceanographic and geographical purposes are as follows:

R packages	Python libraries
• OceanView	• numpy
• marelac	• matplotlib
• oce	• pandas
• ocedata	• basemap
• oceanmap	• scipy
• maps	• netcdf4

Continued

R packages	Python libraries
• mapdata	• iris
• mapproj	• cartopy
• rworldmap	• gsw
• rworldxtra	• seawater
• ggmap	• oceans
• rgdal	• rgdal
• geosphere	• geos
• OpenStreetMap	• shapely
• rgeos	• ctd
• maptools	• proj4
• marmap	• pyproj
• GISTools	• pyshp
• PBSmapping	• geopandas
• spatialkernel	• geoplot
• geoR	• GeoRasters
• automap	• gridded
• RNetCDF	• seabird

In this book, R examples were prepared with R version > 3 and tested with latest R 3.5.1 version.

Python examples were prepared and tested with package versions in Miniconda Python Environment.

A simple conda environment can be created, and necessary packages can be installed from command line/terminal by typing:

Create python environment

```
conda create -n env_name python=2
```

Activate new environment

```
conda activate env_name
```

Install packages

```
conda install -n env_name numpy=1.11.3 matplotlib=1.5.3
pandas=0.17.0 basemap=1.0.7 scipy=0.19.1 netcdf4=1.2.4
cartopy=0.14.3 gsw=3.0.6 seawater=3.3.4 oceans=0.4.0
netcdftime iris=1.9.2 geopy=1.11.0 cmocean
```

Install seabird package with pip

```
pip install seabird
```

References

[1] B.S. Everitt, T. Hothorn, A Handbook of Statistical Analyses Using R, second ed., CRC Press, Florida, 2010.
[2] R-Project Website, About R, n.d. https://www.r-project.org/about.html. Accessed 01 September 2017.
[3] A. Boschetti, L. Massaron, Python Data Science Essentials, first ed., Pack Publishing, Birmingham, 2015.
[4] Python Documentation, Foreword for "Programming Python" n.d. (1st ed.), http://legacy.python.org/doc/essays/foreword/. Accessed 01 September 2017.
[5] Python Documentation, n.d. Modules, https://docs.python.org/2.7/tutorial/modules.html. Accessed 01 September 2017.
[6] Python Documentation, n.d. Packages, https://docs.python.org/2.7/tutorial/modules.html#packages. Accessed 01 September 2017.

CHAPTER 2

Data import and export in R and Python

2.1 Object types in R

Vectors, matrices, arrays, data frames, lists, factors, and functions are object types used to store data in R [1]. Their uses are explained in the following examples.

2.1.1 Vectors

c is a generic function that combines its arguments as a vector.
For details run:

```
?c
```

Examples:
Numeric vector

```
vector1 <- c(-1.5,-1,0,1,1.5)
vector1
[1] -1.5 -1.0 0.0 1.0 1.5
class(vector1)
[1] "numeric"
```

Character vector

```
vector2 <- c("Train","Bus","Taxi","Subway","Tramway")
vector2
[1] "Train" "Bus"  "Taxi" "Subway" "Tramway"
class(vector2)
[1] "character"
```

R and Python for Oceanographers
https://doi.org/10.1016/B978-0-12-813491-7.00002-6

Logical vector

```
vector3 <- c(TRUE,FALSE,TRUE)

vector3
[1] TRUE FALSE TRUE

class(vector3)
[1] "logical"
```

Date vector (defined as character vector and transformed to date)

```
vector4 <- as.Date(c("2003-10-05","2003-10-10","2003-10-15",
"2003-10-20","2003-10-25"))

vector4
[1]"2003-10-05" "2003-10-10" "2003-10-15" "2003-10-20" "2003-10-25"

class(vector4)
[1] "Date"
```

2.1.2 Matrices

Matrix is a function that creates a matrix from the given set of values.

In R, a matrix is the special case of a two-dimensional array.

In a matrix, all columns must have the same class (numeric, character, or logical) and the same length.

Usage of a matrix function:

```
matrix(vector, nrow=r, ncol=c, byrow=FALSE, dimnames=list(char_
vector_rownames, char_vector_colnames))
```

For details, run:

```
?matrix
```

Examples:

create a 2 x 2 numeric matrix

```
mat1 <-matrix(1:4, nrow=2,ncol=2)
```

create a 2 x 2 numeric matrix with row and column names, data is filled by rows.

```
data_vector <- c(1:4)
row_names <- c("R1","R2")
col_names <- c("C1","C2")
mat2 <- matrix(data_vector, nrow=2, ncol=2, byrow=TRUE,
dimnames=list(row_names, col_names))
```

2.1.3 Arrays

Array function creates an array with n dimensions (dim) and their names (dimnames).

An array could have one, two, or more dimensions. A two-dimensional array is the same as a matrix.
Usage of an array function:

```
array(data = NA, dim = length(data), dimnames = NULL)
```

For details, run:

```
?array
```

Examples:

```
vector1 <- c(1,2,3,4)
vector2 <- c(5,6,7,8)
vector3 <- c(9,10,11,12)

matrix_name <- c("Matrix")
matrix_names <- c("Matrix 1","Matrix 2")
row_names <- c("R1","R2","R3","R4")
col_names1 <- c("C1")
col_names2 <- c("C1","C2")
col_names3 <- c("C1","C2","C3")

array1 <- array(vector1, dim=c(4,1,1),
dimnames=list(row_names,col_names1,matrix_name))

array2 <- array(c(vector1,vector2), dim=c(4,2,1),
dimnames=list(row_names,col_names2,matrix_name))

array3 <- array(c(vector1,vector2,vector3), dim=c(4,3,1),
dimnames=list(row_names,col_names3,matrix_name))

array4 <- array(c(vector1,vector2,vector3), dim=c(4,3,2),
dimnames=list(row_names,col_names3,matrix_names))
```

2.1.4 Data frames

In a data frame, different types of vectors (numeric, character, logical, or date) could be defined.

Lengths of the vectors should be the same, and the vectors are stored in columns.

Usage of data frames:

```
data.frame(..., row.names = NULL, check.rows = FALSE,
        check.names = TRUE, fix.empty.names = TRUE,
        stringsAsFactors = default.stringsAsFactors())
```

For details, run:

```
?data.frame
```

Example:

```
vector1 <- c(-1.5,-1,0,1,1.5)
vector2 <- c("Train","Bus","Taxi","Subway","Tramway")
vector3 <- c(TRUE,FALSE,TRUE,TRUE,FALSE)
vector4 <- as.Date(c("2003-10-05","2003-10-10","2003-10-
15","2003-10-20","2003-10-25"))

data_frame <- data.frame(num=vector1, char=vector2,
log=vector3,date=vector4,stringsAsFactors=FALSE)

names(data_frame) <-
c("numeric","character","logical","date")
```

2.1.5 Lists

In lists, objects are stored in the order they are given. Different types of objects can be stored in lists.

For details, run:

```
?list
```

Examples:

```
vector1 <- c(-1.5,-1,0,1,1.5)
vector2 <- c("Train","Bus","Taxi","Subway","Tramway")
```

```
vector3 <- c(TRUE,FALSE,TRUE)
vector4 <- as.Date(c("2003-10-05","2003-10-10","2003-10-15"))

matrix1 <-matrix(1:4, nrow=2,ncol=2)

list1 <- list(vec1=vector1, vec2=vector2, vec3=vector3,
vec4=vector4, mat=matrix1)

list2 <- list(vec2=vector2, vec4=vector4)

list3 <- c(list1,list2)
```

2.1.6 Factors

Factor is a function used to define a vector as a factor. The orders of variables in a factor can be controlled by levels.
For details, run:

```
?factor
```

Examples:

```
vector1 <- c(5,10,15,20)
factor1 <- factor(vector1)
factor1

factor2 <- factor(vector1, levels=c(20,10,5,15))
factor2

factor3 <- factor(vector1, levels=c(10,5,20,15),
labels=c("august","february","may","november"))
factor3
```

2.1.7 Functions

R has many built-in functions, but in some cases, it is more useful to create user-defined functions for a specific task.
A sample function to find roots of a quadratic equation:

```
quad_root <- function(a, b, c){

    delta <- (b^2) - 4*a*c

    if (delta>0) {
```

Continued

```
        pos_root <- format(((-b) + sqrt(delta)) / (2*a),
digits=2, nsmall=2)
        neg_root <- format(((-b) - sqrt(delta)) / (2*a),
digits=2, nsmall=2)
        print(paste("Positive root: ",pos_root))
        print(paste("Negative root: ",neg_root))
    } else if (delta==0) {
        root <- format((-b) / (2*a), digits=2, nsmall=2)
        print(paste("Root: ",root))
    } else {
        root_real <- format((-b) / (2*a), digits=2,
nsmall=2)
        root_imag <- format(sqrt(abs(delta)) / (2*a),
digits=2, nsmall=2)
        print(paste("For quadratic equation of ",a,"x^2 +",b,
"x +",c,":"))
        print(paste("Roots are complex numbers: ",
root_real, "\U00B1", root_imag,"i"))
        }
}
```

define coefficients of the 2nd order polynomial

```
a <- 5; b <- 4; c <- 3
quad_root(a,b,c)
[1] "For quadratic equation of 5 x^2 + 4 x + 3 :"
[1] "Roots are complex numbers: -0.40 ± 0.66 i"
```

You can also use built-in polyroot function to find roots.

```
polyroot(c(3,4,5))
[1] -0.4+0.663325i -0.4-0.663325i
```

2.2 Data import in R

Data can be imported from many file types such as txt, dat, csv, xls, xlsx, nc (netCDF4), mat (Matlab), and cnv (Seabird). These are some of the file types mostly used in storage of oceanographic data. In the following data importing examples, data files exported in Section 2.3 can be used.

2.2.1 Import from txt, dat, csv and Excel xls, xlsx files

```
data <- read.table("/path/data.txt", header=TRUE, sep="/t")

data <- read.table("/path/data.dat", header=TRUE, sep=";")

data <- read.table("/path/data.csv", header=TRUE, sep=",")

library(xlsx)

data <- read.xlsx("/path/data.xls", sheetIndex=1, header=TRUE)

data <- read.xlsx("/path/data.xlsx", sheetIndex=1, header=TRUE)

data <- read.xlsx("/path/data.xlsx", sheetName="Sheet1",
header=TRUE)
```

2.2.2 Import from netCDF4 file

```
library(ncdf4)
ncin <- nc_open("/path/seasurf_ts.nc")
ncin

lon <- ncvar_get(ncin, "lon")

lat <- ncvar_get(ncin, "lat")

t1 <- ncvar_get(ncin, "temp_val")

s1 <- ncvar_get(ncin, "sal_val")
```

2.2.3 Import from mat files

```
setwd('/path_to_working_directory/')
library(R.matlab)

data <- readMat('data_m5.mat')
str(data)
```

2.2.4 Import from SeaBird cnv files

```
library(oce)
f <- read.ctd("/path/ctd.cnv")
f
summary(f)
plot(f)
f[['temperature']]
```

Data in this example was retrieved from GitHub page of oce package:
"https://github.com/dankelley/oce/blob/develop/tests/testthat/local_data/ctd.cnv"

2.2.5 Import data from online databases

Import tab delimited text data from LOBO-0010 Northwest Arm, Halifax, Canada

This example demonstrates reading data in text format from online database of Land/Ocean Biogeochemical Observatory (LOBO) system measuring biogeochemical parameters with chemical and physical sensors at Northwest Arm, Halifax, and Canada (URL: http://lobo.satlantic.com/).

```
url <- 'http://lobo.satlantic.com/cgi-data/nph-data.cgi?
min_date=20101225&max_date=20101230&y=temperature,salinity'
colnames <- c('DATE_TIME','TEMPERATURE','SALINITY')
tmpFile <- tempfile()
download.file(url, destfile = tmpFile, method = "auto")
url.data <- read.table(tmpFile, header=FALSE, sep='\t',
col.names=colnames, skip=1)
print(url.data)
```

2.2.6 Import csv.gz file from WOA13 V2

This example demonstrates reading data in compressed csv format from the online database of World Ocean Atlas 2013 V2 (WOA13 V2) maintained by NOAA (URL: https://www.nodc.noaa.gov/OC5/woa13/). WOA13 V2 is a set of climatological data with 1° grid resolution and provides annual, seasonal, and monthly in situ temperature, salinity, dissolved oxygen, Apparent Oxygen Utilization (AOU), percent oxygen saturation, phosphate, silicate, and nitrate data at standard depth levels for the World Ocean.

```
url <-
'http://data.nodc.noaa.gov/woa/WOA13/DATAv2/temperature/
csv/decav/1.00/woa13_decav_t07mn01v2.csv.gz'

colnames <-
c('LATITUDE','LONGITUDE','0','5','10','15','20','25','30','35',
'40','45','50','55','60','65','70','75','80','85','90','95',
'100','125','150','175','200','225','250','275','300','325',
'350','375','400','425','450','475','500','550','600','650',
'700','750','800','850','900','950','1000','1050','1100',
'1150','1200','1250','1300','1350','1400','1450','1500')

tmpFile <- tempfile()

download.file(url, destfile = tmpFile, method = "auto")

url.data <- read.csv(tmpFile, header=FALSE, sep=',',
col.names=colnames, skip=2)

datasub <- subset(url.data, LATITUDE=="36.5" & LONGITUDE > 0
& LONGITUDE < 30, select=c(LATITUDE,LONGITUDE,X0,X5,X10))

print(datasub)
```

2.3 Data export in R

Raw or processed data in R can be exported as file types, such as txt, dat, csv, xls, xlsx, nc (netCDF4), and mat (Matlab).

2.3.1 Export as txt, dat, csv and Excel xls, xlsx files

Import data

```
data <- read.table("/path/data.txt ", header=TRUE, sep="\t")
```

Export data

```
write.table(data, "/path/data.txt", sep="\t", quote=FALSE,
dec=".", row.names=FALSE, col.names=TRUE)

write.table(data, "/path/data.dat", sep=";", quote=FALSE,
dec=".", row.names=FALSE, col.names=TRUE)
```

Continued

```
write.table(data, "/path/data.csv", sep=",", quote=FALSE,
dec=".", row.names=FALSE, col.names=TRUE)

library(xlsx)

write.xlsx(data, "/path/data.xls", sheetName="Sheet1",
col.names=TRUE, row.names=FALSE, append=FALSE, showNA=TRUE,
password=NULL)

write.xlsx(data, "/path/data.xlsx", sheetName="Sheet1",
col.names=TRUE, row.names=FALSE, append=FALSE, showNA=TRUE,
password=NULL)
```

2.3.2 Export as netCDF4 file

```
library(ncdf4)

setwd("/path_to_working_directory/")
```

generate lons, lats and set time

```
lon <- as.array(seq(26.59,26.72,0.0005))
nlon <- dim(lon)

lat <- as.array(seq(38.57,38.67,0.0005))
nlat <- dim(lat)

temp <- as.array(rnorm(nlon*nlat, mean=18, sd=2.5))
ntemp <- dim(temp)
tunit <- "Celcius"

sal <- as.array(rnorm(nlon*nlat, mean=38, sd=1.2))
nsal <- dim(sal)
sunit <- "psu"
```

nlon * nlat * nt array

```
temp_array <- array(temp, dim=c(nlon,nlat))
sal_array <- array(sal, dim=c(nlon,nlat))
```

define dimensions

```
londim <- ncdim_def("lon","degrees_east",as.double(lon))
latdim <- ncdim_def("lat","degrees_north",as.double(lat))
tempdim <- ncdim_def("temp",tunit,as.double(temp))
saldim <- ncdim_def("sal",sunit,as.double(sal))
```

define variables

```
fillvalue <- 1e32

dlname <- "sea surface temperature"

temp_def <-
ncvar_def("temp_val","deg_C",list(londim,latdim),fillvalue,
dlname,prec="single")

dlname <- "sea surface salinity"

sal_def <-
ncvar_def("sal_val","psu",list(londim,latdim), fillvalue,
dlname,prec="single")
```

create netCDF file and add arrays

```
ncfname <- "seasurf_ts.nc"
ncout <-
nc_create(ncfname,list(temp_def,sal_def ),force_v4=T)
```

add variables

```
ncvar_put(ncout,temp_def,temp_array)
ncvar_put(ncout,sal_def,sal_array)
```

add additional attributes into dimension and data variables

```
ncatt_put(ncout,"lon","axis","X") #,verbose=FALSE)
#,definemode=FALSE)
ncatt_put(ncout,"lat","axis","Y")
```

add global attributes

```
title <- paste("Sea Surface Temperature and Salinity")
institution <- paste("DEU, IMST")
source <- paste("Project 1, Survey 1")
```

Continued

```
history <- paste("H.A.", date(), sep=",")
references <- paste("")
conventions <- paste("")

ncatt_put(ncout,0,"title",title)
ncatt_put(ncout,0,"institution",institution)
ncatt_put(ncout,0,"source",source)
ncatt_put(ncout,0,"references",references)
ncatt_put(ncout,0,"history",history)
ncatt_put(ncout,0,"Conventions",conventions)
```

save and close the file

```
nc_close(ncout)
```

2.3.3 Export as Matlab mat files

```
setwd(/path_to_working_directory/')

library(R.matlab)

lon <- as.array(seq(26.59,26.72,0.0005))
lat <- as.array(seq(38.57,38.67,0.0005))
temp <- as.array(rnorm(dim(lon)*dim(lat), mean=18, sd=2.5))
sal <- as.array(rnorm(dim(lon)*dim(lat), mean=38, sd=1.2))

A <- matrix(1:30, ncol=3)
B <- as.matrix(1:20)
C <- array(1:15, dim=c(2,2,3))

writeMat("data_m5.mat", lon=lon, lat=lat, temp=temp, sal=sal,
A=A, B=B, C=C)
```

2.4 Object types in Python

In Python, numeric and character data are stored as lists, tuples, dictionaries, sets, and functions. It is also possible to store numeric data as NumPy arrays, Pandas series, and Pandas data frames (Table 2.1) [2,3].

Table 2.1 Standard data and object types in Python

Object type	Example	Properties
Lists	[1, 12, 2.46]	Mutable sequences, used with square brackets.
Tuples	(3, 4.5, 'unit', 'D'), tuple('extra')	Immutable sequences, used with parenthesis.
Dictionaries	{'Jack': 'train', 'Sally': 'subway'}	Data stored as key: value pair. Used with curly braces.
Sets	{'r', 'u', 'n'}, set('run')	Collections of unique elements.

2.4.1 Arrays in Numpy package

Create an array

Usage of array function:

```
numpy.array(object, dtype = None, copy = True, order = None,
subok = False, ndmin = 0)
```

For more details, run:

```
help(np.array)
```

Example:

```
array1 = np.array([1,2,3])
print array1
array2 = np.array([1,2,3],[4,5,6])
print array2
```

Create array from existing data

Usage of asarray function:

```
numpy.asarray(a, dtype = None, order = None)
```

For more details, run:

```
help(np.asarray)
```

Example:

```
import numpy as np

list1 = [1,2,3,4]
list2 = [5,6,7,8]
list3 = [9,10,11,12]

array1 = np.asarray([list1, list2, list3])

print array1

array2 = np.transpose(array1)

print array2
```

2.4.2 Series and data frame objects in Pandas package

Pandas Series

Pandas Series is a one-dimensional array with axis labels, and it can store different types of data (integer, string, other python objects, etc.) [2].
Usage of Series function:

```
pd.Series(data, index, dtype, copy)
```

For more details, run:

```
help(pd.Series)
```

Example:

```
import numpy as np
import pandas as pd

array = np.array([1,2,3])

print array

series = pd.Series(array)

print series

data = [1,2,3]

series2 = pd.Series(data)

print series2
```

Pandas Data Frames

Pandas Data Frame has a two-dimensional data structure and stores data in a table-like format within rows and columns. Different data types can be stored in different columns; mutable object, rows, and columns can be labeled, and arithmetic operations can be performed on rows and columns. Pandas Data Frames can be created from objects like lists, sets, dictionaries, and NumPy arrays.
Usage of DataFrame function:

```
pd.DataFrame(data, index, columns, dtype, copy)
```

```
import numpy as np
import pandas as pd

list1 = [1,2,3,4]
list2 = [5,6,7,8]
list3 = [9,10,11,12]

array1 = np.asarray([list1, list2, list3])
array2 = np.transpose(array1)

data_frame = pd.DataFrame(array2,columns=['Col1','Col2','Col3'])

data = [[1,5,9],[2,6,10],[3,7,11],[4,8,12]]

data_frame2 = pd.DataFrame(data,columns=['Col1','Col2','Col3'])
```

2.4.3 User-defined functions in Python

Python has many built-in functions, but in some cases, it is useful to create user-defined functions for a specific task.
A sample function to find roots of a quadratic equation:

```
import math

def quad_root(a,b,c):

    delta = b**2 - 4*a*c

    if delta > 0:
        pos_root = ((-1.*b)+math.sqrt(delta))/(2*a)
        neg_root = ((-1.*b)-math.sqrt(delta))/(2*a)
        print "Positive root: %.2f" % (pos_root)
        print "Negative root: %.2f" % (neg_root)
```

Continued

```
    elif delta == 0:
        root = (-1.*b) / (2*a)
        print "Root: %.2f" % (root)
    else:
        root_real = (-1.*b) / (2*a)
        root_imag = math.sqrt(math.fabs(delta)) / (2*a)
        print "For quadratic equation of %d x^2 + %d x + %d
:" % (a,b,c)
        print "Roots are complex numbers: %.2f +- %.2f i" %
(root_real,root_imag)
```

define coefficients of the 2nd order polynomial

```
a = 5
b = 4
c = 3

quad_root(a,b,c)
```

You can also use built-in numpy.roots function to find roots:

```
import numpy as np

coeffs = [a,b,c]

roots = np.roots(coeffs)
```

2.5 Data import in Python

In Python, data can be imported from many file types, such as txt, dat, csv, xls, xlsx, nc (netCDF4), mat (Matlab), and cnv (Seabird). In the following data importing examples, data files in Section 2.6 can be used.

2.5.1 Import from txt, dat, csv, and Excel xls, xlsx files with Numpy loadtxt

```
import numpy as np
import os
os.chdir(/path_to_working_directory/')

depth,do,phos,tnox,si = np.loadtxt('data.csv', unpack=True,
usecols=None, skiprows=1, delimiter=',')
```

```
depth,do,phos,tnox,si = np.loadtxt('data.txt', unpack=True,
usecols=None, skiprows=1, delimiter='\t')

depth,do,phos,tnox,si = np.loadtxt('data.dat', unpack=True,
usecols=None, skiprows=1, delimiter=';')
```

2.5.2 Import from Excel xls files with Pandas read_excel

```
import pandas as pd
import numpy as np
import os
os.chdir(/path_to_working_directory/')
np.set_printoptions(precision=4)
pdata = pd.read_excel('data.xls')
ndata = np.array(pdata, 'g')
ndata

depth = np.take(ndata, 0, axis=1)
do = np.take(ndata, 1, axis=1)
phos = np.take(ndata, 2, axis=1)
tnox = np.take(ndata, 3, axis=1)
si = np.take(ndata, 4, axis=1)
```

2.5.3 Import from Excel xlsx files with Pandas read_excel

```
import pandas as pd
import numpy as np
import os
os.chdir(/path_to_working_directory/')
np.set_printoptions(precision=4)
pdata = pd.read_excel('data.xlsx')
ndata = np.array(pdata, 'g')
ndata

depth = np.take(ndata, 0, axis=1)
do = np.take(ndata, 1, axis=1)
phos = np.take(ndata, 2, axis=1)
tnox = np.take(ndata, 3, axis=1)
si = np.take(ndata, 4, axis=1)
```

2.5.4 Import from netCDF4 file

```
import netCDF4 as nc
```

open a local NetCDF file or remote data URL

```
file = '/path/py_ss_ts.nc'
ncd = nc.Dataset(file, 'r')
```

Print variables, dimensions, units and general descriptions

```
print ncd
```

Print attributes for all variables

```
print ncd.variables
```

Print all variable names

```
print ncd.variables.keys()
```

Print all attributes of a specific variable

```
print ncd.variables['Temperature']
tvar = ncd.variables['Temperature']
```

Print attribute names of a specific variable

```
print tvar.ncattrs()
```

Print units of a variable

```
print tvar.getncattr('units')
```

Copy values from a specific variable to a newly defined variable

```
tempv = ncd.variables['Temperature'][:]
```

2.5.5 Import from Matlab mat files

```
import scipy.io

file 1 = '/path/data_m5.mat'

mat1 = scipy.io.loadmat(file1, mat_dtype=True)

vars1 = scipy.io.whosmat(file1)

mat1.keys()

temp = mat1['temp']

type(temp)
```

2.5.6 Import from Seabird cnv file

```
import os
os.chdir('/path_to_working_directory/')

from seabird.cnv import fCNV

profile = fCNV('dPIRX003.cnv')
profile.attributes

print "The profile coordinates is latitude: %.4f, and
longitude: %.4f" % \
    (profile.attributes['LATITUDE'],
profile.attributes['LONGITUDE'])

print profile.keys()
print profile['TEMP'].mean(), profile['TEMP'].std()

from matplotlib import pyplot as plt

plt.plot(profile['TEMP'], profile['PRES'],'b')
plt.plot(profile['TEMP2'], profile['PRES'],'g')
plt.gca().invert_yaxis()
plt.xlabel('temperature')
plt.ylabel('pressure')
plt.title(profile.attributes['filename'])
plt.show()
```

2.5.7 Import data from online databases

Import tab delimited text data from LOBO-0010 Northwest Arm, Halifax, Canada

```
import pandas as pd

url = 'http://lobo.satlantic.com/cgi-data/nph-data.cgi?
min_date=20101225&max_date=20101230&y=temperature,salinity'

colnames = ['DATE_TIME','TEMPERATURE','SALINITY']

data = pd.read_csv(url, delimiter='\t', compression=None,
error_bad_lines=False, warn_bad_lines=False, skiprows=[0],
names=colnames)

print data
```

Import all data in a csv file

```
import pandas as pd

url =
'https://data.nodc.noaa.gov/woa/WOA13/DATAv2/temperature/csv/
decav/1.00/woa13_decav_t07mn01v2.csv.gz'

colnames =
['LATITUDE','LONGITUDE','0','5','10','15','20','25','30','35',
'40','45','50','55','60','65','70','75','80', '85','90','95','
100','125','150','175','200','225','250','275','300','325',
'350','375','400','425','450','475','500','550','600','650',
'700','750','800','850','900','950','1000', '1050','1100',
'1150','1200','1250','1300','1350','1400','1450','1500']

data = pd.read_csv(url, delimiter=',', compression='gzip',
error_bad_lines=False, warn_bad_lines=False, skiprows=[0,1],
names=colnames)

print data
```

Import block of data at defined ranges of rows and columns

It is possible to import data as a block of rows and columns by defining use-cols and nrows parameters in read_csv function.

```
import pandas as pd
url =
'https://data.nodc.noaa.gov/woa/WOA13/DATAv2/temperature/csv/
decav/1.00/woa13_decav_t07mn01v2.csv.gz'
colnames =
['LATITUDE','LONGITUDE','0','5','10','15','20', '25','30',
'35','40','45','50','55','60','65','70','75','80','85','90',
'95','100','125','150','175','200','225','250','275','300',
'325','350','375','400','425','450','475','500','550','600',
'650','700','750','800','850','900','950','1000','1050','1100',
'1150','1200','1250','1300','1350','1400', '1450','1500']
data = pd.read_csv(url, delimiter=',', compression='gzip',
error_bad_lines=False, warn_bad_lines=False,
usecols=range(0,8,1), nrows=100, skiprows=20000,
names=colnames)
print data
```

Subset data based on multiple criteria using column values

```
import pandas as pd
url =
'https://data.nodc.noaa.gov/woa/WOA13/DATAv2/temperature/csv/
decav/1.00/woa13_decav_t07mn01v2.csv.gz'
colnames =
['LATITUDE','LONGITUDE','0','5','10','15','20','25','30',
'35','40','45', '50','55','60', '65','70','75', '80','85','90',
'95','100','125','150','175','200','225','250','275','300',
'325','350','375','400','425','450','475','500','550','600',
'650','700','750','800','850','900','950','1000','1050','1100',
'1150','1200','1250','1300','1350','1400','1450','1500']
data = pd.read_csv(url, delimiter=',', compression='gzip',
error_bad_lines=False, warn_bad_lines=False, skiprows=[0,1],
names=colnames)
```

Subset Example 1:

```
data[['LATITUDE', 'LONGITUDE','0','5','10']].query('LATITUDE
== 36.5 & 0 <= LONGITUDE < 30')
```

Subset Example 2:

```
data[data['LATITUDE'] == 36.5][data['LONGITUDE'] >= 0]
[data['LONGITUDE'] < 30][['LATITUDE','LONGITUDE','0','5','10']]
```

2.6 Data export in Python

Raw or processed data in Python can be exported as file types such as txt, dat, csv, xls, xlsx, nc (netCDF4) and mat (Matlab).

2.6.1 Export as txt, dat, and csv files with Numpy savetxt

```
import numpy as np
import os
os.chdir('/path_to_working_directory/')

temp = np.random.normal(loc=18, scale=2.5, size=20)
sal = np.random.normal(loc=38, scale=1.2, size=20)

txt_header = 'temp\tsal'
dat_header = 'temp;sal'
csv_header = 'temp,sal'

np.savetxt('data.txt', zip(temp,sal), delimiter='\t',
header=txt_header, fmt="%.2f")

np.savetxt('data.dat', zip(temp,sal), delimiter=';',
header=dat_header, fmt="%.2f")

np.savetxt('data.csv', zip(temp,sal), delimiter=',',
header=csv_header, fmt="%.2f")
```

2.6.2 Export as txt, dat and csv files with Pandas DataFrame.to_csv

```
import numpy as np
import pandas as pd
import os
os.chdir('/path_to_working_directory/')

temp = np.random.normal(loc=18, scale=2.5, size=20)
sal = np.random.normal(loc=38, scale=1.2, size=20)
```

```
dataset = pd.DataFrame({'temp':temp,'sal':sal})
dataset.to_csv('data.txt', sep='\t', index=False)
dataset.to_csv('data.dat', sep=';', index=False)
dataset.to_csv('data.csv', sep=',', index=False)
```

2.6.3 Export as xls, xlsx with Pandas ExcelWriter

```
import numpy as np
import pandas as pd
from pandas import ExcelWriter
import os
os.chdir('/path_to_working_directory/')

temp = np.random.normal(loc=18, scale=2.5, size=20)
sal = np.random.normal(loc=38, scale=1.2, size=20)

dataset = pd.DataFrame({'temp':temp,'sal':sal})

xls_write = ExcelWriter('data.xls')
dataset.to_excel(xls_write,'Sheet1',index=False)
xls_write.save()

xlsx_write = ExcelWriter('data.xlsx')
dataset.to_excel(xlsx_write,'Sheet1',index=False)
xlsx_write.save()
```

2.6.4 Export as netCDF4 files

```
from datetime import datetime
import numpy as np
import netCDF4 as nc
```

Prepare dataset

```
lon = np.arange(26.59,26.72,0.0005)
lat = np.arange(38.57,38.67,0.0005)
temp = np.random.normal(loc=18, scale=2.5,
size=len(lon)*len(lat))
sal = np.random.normal(loc=38, scale=1.2,
size=len(lon)*len(lat))
```

Create netCDF4 dataset

```
file = nc.Dataset('/path/py_ss_ts.nc','w', format='NETCDF4')
```

Create datagroup

```
#datagr = file.createGroup('data')
```

Create dimenions of variables

```
file.createDimension('lond', len(lon))
file.createDimension('latd', len(lat))
file.createDimension('tempd', len(temp))
file.createDimension('sald', len(sal))
```

Create variables in netCDF4 file

```
lonv = file.createVariable('Longitude', 'f4', 'lond')
latv = file.createVariable('Latitude', 'f4', 'latd')
tempv = file.createVariable('Temperature', 'f4', 'tempd')
salv = file.createVariable('Salinity', 'f4', 'sald')
```

Link data into variables

```
lonv[:] = lon
latv[:] = lat
tempv[:] = temp
salv[:] = sal
```

#Add global attributes

```
today = datetime.today()
file.description = "Sea Surface Temperature and Salinity"
file.history = "H.A. " + today.strftime("%d/%m/%y")
file.institution = "DEU, IMST"
file.source = "Project 1, Survey 1"
file.references = ""
file.conventions = ""
```

#Add local attributes to variable instances

```
lonv.units = 'degrees east'
latv.units = 'degrees north'
tempv.units = 'Celcius'
salv.units = 'psu'

file.close()
```

2.6.5 Export as Matlab mat files

```
import os
os.chdir('/path_to_working_directory/)
import scipy.io
import numpy as np

lon = np.arange(26.59,26.72,0.0005)
lat = np.arange(38.57,38.67,0.0005)
temp = np.random.normal(loc=18, scale=2.5,
size=len(lon)*len(lat))
sal = np.random.normal(loc=38, scale=1.2,
size=len(lon)*len(lat))

scipy.io.savemat('np_array.mat', {'lon':lon, 'lat':lat,
'temp':temp, 'sal':sal})
```

References

[1] R Manuals, An Introduction to R – Graphical procedures, https://cran.r-project.org/doc/manuals/r-release/R-intro.html. Accessed 19 July 2018.
[2] Pandas Documentation, http://pandas.pydata.org/pandas-docs/stable/. Accessed 19 July 2018.
[3] NumPy Documentation, https://docs.scipy.org/doc/. Accessed 1 January 2018.

CHAPTER 3

Plotting

3.1 Plots in R

In R, there are many possibilities with a wide number of built-in plotting functions to visualize your data. Plotting functions are classified as high-level and low-level functions. High-level functions are used to create new plots with automatically generated axes, titles, labels, and other plot properties. Low-level functions provide the ability to add more information to an existing plot, such as texts, lines, and points. Besides these functions, it is also possible to customize plots by changing graphical parameters. In general, data and graphical parameters are introduced as arguments in these functions to generate a plot.

3.1.1 High-level plotting functions

In R, high-level plotting functions are plot(), qqplot(), qqnorm(), qqline(), hist(), dotchart(), image(), contour() and persp(). From these, plot() is a generic function that produces plots based on the data type introduced in the first argument [1].

plot() function

Many types of data can be introduced to plot() function, and it automatically produces the plot.
To produce a scatterplot y against x:

```
plot(x, y)  # x and y are numeric vectors.
plot(xy)    # xy is a list two elements or a two column matrix
```

Time series plot:

```
plot(x)  # x is a numeric vector including index values
```

Barplot or boxplot from a factor object and a numeric vector:

```
plot(f)    # f is a factor object, produces barplot
plot(f, y) # y is a numeric vector, produces boxplot
```

R and Python for Oceanographers
https://doi.org/10.1016/B978-0-12-813491-7.00003-8

Other high-level plotting functions

Plots for checking distribution of data:

```
qqnorm(x)    # x is a numeric vector, produces plot of x against
its Normal scores

qqline(x)    # x is a numeric vector,
produces a line passing through the distribution and data
quartiles

qqplot(x, y) # x and y are numeric vectors, compares
distributions of x against y
```

Histogram plots:

```
hist(x)                    # x is a numeric vector

hist(x, breaks=b, ...)     # breaks: manually enter
breakpoints in data
```

Dotchart plot

```
dotchart(x, ...)   # x is a numeric vector, y-axis is used for
labelling of data
```

X-Y-Z plots

```
image(x, y, z, ...)     # draws rectangular grids from x,y and
colors them with z values

contour(x, y, z, ...)   # draws contour lines

persp(x, y, z, ...)     # draws 3D surface
```

Arguments to use with high-level plotting functions

Arguments are used to control the properties of the plot. Some examples to arguments used with high-level plotting functions are as follows:

```
add=TRUE or FALSE    # If TRUE, superimposes the plot on to an
existing plot

axes=TRUE or FALSE   # If FALSE, suppresses the plotting of axes

log="x" or "y" or "xy"   # Converts x, y or both axes to
logarithmic scale

type="p" or "l" or "b" or "o" or "h" or "s" or "n"   # p: points
(default), l: lines, b: points connected with lines, o: points
```

```
overlaid by lines, h: vertical lines from points to the zero
axis, s: step-function plot, n: data not presented but axes are
drawn from data.

xlab="x axis label"          # String

ylab="y axis label"          # String

main="plot title"            # String

sub="sub-title for plot"     # String
```

Low-level plotting functions

In some cases, customized plots are needed instead of standard ones. Low-level plotting functions can be used for this purpose to add more information to an existing plot such as texts, lines and points.

Some examples of low-level plotting functions are:

```
points(x, y)                 # Adds points

lines(x, y)                  # Adds lines

text(x, y, labels, ...)  # Adds texts defined with label
(integer or character) argument

abline(a, b)                 # Adds a straight line (y=bx+a),
where b is slope and a is intercept

abline(h=y)                  # Adds horizontal line at y

abline(v=x)                  # Adds vertical line at x

abline(lm.obj)               # Adds a line from a linear model

polygon(x, y, ...)           # Adds a polygon

legend(x, y, legend, ...)    # Adds a legend at a given position

title(main, sub)             # Adds a main title and sub-title on
top of the figure

axis(side, ...)              # Adds an axis to specified axis side
of the plot
```

3.1.2 Graphical parameters

In R, it is possible to produce well-designed graphics, especially for presentations and publications, by using graphical parameters instead of using default functions. With these parameters, you can customize graph elements like line type, line style, line color, font type, and sub-plot arrangements.

These parameters should be set by the user for each graphics device. Parameters can be set permanently to apply on all graphics functions, or temporarily to apply only on a single function.

par() function

The par() function includes a set of graphical parameters and it is used to customize the current graphics device.

```
par()         # Displays the list of all graphical parameters with
their current values

par("arg")      # Displays the value of selected argument
or values in a list, c("arg1","arg2")

par(c("arg1","arg2"))          # Displays values of parameters
in a list

par(font=1, family="sans")      # Sets new values for
given parameters
```

In some cases, it is useful to save and restore initial graphics parameters. This can be done by saving some or all graphics' parameters before using graphics' functions and restoring them at the end.

```
init <- par(font=1, family="sans")   # plotting functions
including graphical parameters

par(init)   # Return to initial font and family values
```

To restore all graphical parameters:

```
init <- par(no.readonly=TRUE)   # plotting functions including
graphical parameters

par(init)   # Return to initial values
```

3.1.3 Graphical parameters as arguments in graphics functions

Graphical parameters may be assigned as arguments in graphical functions. This option lets users change only graphical parameters of that function.

```
plot(x, y, font=2)   # Gives a plot with bold fonts and only
affects this function
```

However, sometimes it is necessary to set or reset graphical parameters with par() function.

3.1.4 Most frequently used graphical parameters

A list of most frequently used graphical parameters is given below. More detailed information could be found from the help page of related graphics function.

```
pch="."  # defines point type, accepts "character" or an
integer between 0 and 25
```

A symbol list to use as points in plots can be visualized with below script:

```
plot.new()

legend("center", as.character(0:25), pch = 0:25, ncol=5,
cex=1.5)
```

Alternatively, a character list from current font can also be used by setting an integer value between 32 and 255 for pch. A script of these font characters can be visualized as follows:

```
plot.new()

legend("center", as.character(32:255), pch = 32:255,
ncol=10, cex=0.9)

lty=1          # defines line type, accepts an integer from 0
to 6 or line name
```

A list of line types can be visualized as below:

```
plot.new()

lty.name <-
c("blank","solid","dashed","dotted","dotdash","longdash",
"twodash")

legend("center", lty.name, lty=0:6, lwd=2, cex=1.5)

lwd=1  # defines line width, accepts a positive number

col=1  # defines color for points, lines, text,
filled regions and images
```

Colors can be set by using:

- color names (can be displayed by using colors() function)
- numbers from 1 to 657
- hexadecimal codes

Color palettes are another option to use colors in your plot. From built-in library(grDevices), pre-defined color palettes are:

```
cm.colors(n)

heat.colors(n)

rainbow(n)

terrain.colors(n)

topo.colors(n)

palette()            # contains 8 pre-defined colors,
colors can be customized, "default" resets to initial colors

colorRamp()          # creates given number of colors from
initial colors defined between [0,1]

colorRampsPalette()  # creates custom color palette with n
colors from initially defined colors
```

Other pre-defined color palette options can be found in other packages like colorRamps, RcolorBrewer, and colorspace packages:

```
col.axis         # defines axis color
col.lab          # defines label color
col.main         defines main title color
col.sub          # defines sub-title color

font=3           # defines font type: 1 for plain text, 2 for
bold, 3 for italic, 4 for bold italic and 5 for symbol

font.axis        # defines axis tick label font
font.lab         # defines axis label font
font.main        # defines main title font
font.sub         # defines sub-title font

adj=-0.2         # defines justification of text: 0 for left
justify, 1 for right justify, 0.5 for centering horizontally
-0.2 means to leave 20% gap from the left margin

cex=1.1          # Adjust text font size, default is 1
cex.axis         # Adjust axis tick label font size
cex.lab          # Adjust axis label font size
cex.main         # Adjust main title font size
cex.sub          # Adjust sub-title font size

lab=c(6, 8, 10)  # First two variables adjusts tick
intervals for x and y axes, respectively, and third variable
defines number of ticks per axes
```

```
las=0               # Adjusts orientation of axis labels: 0 for
parallel to axis, 1 for horizontal layout, 2 for perpendicular
to axis

xaxs="r"

yaxs="i"                    # Adjusts axis style, "r" (default),
"i", "e", "s", "d" values can be used

mai=c(1.5, 1, 1, 0.5)    # Adjusts bottom, left, top and
right margin sizes, unit in inches

mar=c(5, 3, 3, 2)         # Adjusts bottom, left, top and
right margin sizes, unit in text lines
```

3.1.5 Plotting multiple figures

In R, it is possible to create figures with n x m array of sub-plots. It is possible to control each sub-plot and main figure margins with the following arguments:

```
mfcol=c(2, 3)
mfrow=c(3, 4)

omi=c(1.5, 1, 2, 0.5) Adjusts outer margin sizes from bottom,
left, top and right, unit in inches

oma=c(2, 1, 2, 1) Adjusts outer margin sizes from bottom, left,
top and right, unit in text lines
```

If outer margins are defined, it is also possible to add text to outer margin area by setting outer=TRUE within mtext() command.

3.1.6 Device drivers

In R, graphics devices are used to translate plotting commands into a form that is recognized by display or printing devices. Based on the needs of the user, graphics devices can be used to display the figure or to save it in the preferred graphics format. This is usually performed by calling the graphics device command before plotting commands. However, if no graphics device is opened, then calling a high-level plot function will open a default one.

Graphics device commands used to display plots are as follows:

```
dev.new()  # Opens a new graphics window

X11()       # Opens graphics window in Unix-like systems

quartz()    # Opens a graphics window in Mac OSX

windows()   # Opens graphics window in Windows
```

To save the plot, the following graphics devices can be used:

```
postscript()    # Saves the graphics in Postscript (ps) format
pdf()           # Saves the graphics in Printable Document File
(pdf ) format
bitmap()        # Saves the graphics in Bitmap (bmp) format
svg()           # Saves the graphics in Scalable Vector Graphics
(svg) format
png()           # Saves the graphics in Portable Network Graphics
(png) format
jpeg()          # Saves the graphics in JPEG format
tiff()          # Saves the graphics in TIFF format
savePlot()      # Saves the graphics in defined format
dev.off()       # Closes the current or specified graphics
device, it is used at the end of plotting commands
```

3.1.7 Example plots in R

A point plot for a non-linear function (Fig. 3.1):

```
x <- rnorm(100, mean=1.0, sd=0.25)
y <- x^2+x
legend <- expression(y=x^2+x)
tiff(filename="figure.tiff", width=15, height=12, unit="cm",
     res=300, pointsize=12, compression="lzw", bg="white")
par(mar=c(4,5,3,5), las=1)
plot(x, y, type="p", col="blue", pch=16,
     main="Main Title", xlab="x axis label",
     ylab="y axis label")
legend("topleft", legend=legend,
     lty=0, pch=16, col="blue",
     ncol=1, bty="o", cex=1,
     text.col="black", box.col="gray",
     inset=0.01)
dev.off()
```

A two axis point plot for a linear and a non-linear function (Fig. 3.2):

```
x <- rnorm(100, mean=1.0, sd=0.25)
y1 <- 2*x+5
y2 <- -x^3+x
```

```
legend1 <- expression(y1=2*x+5)
legend2 <- expression(y2=-x^3+x)

tiff(filename="fig4.tiff", width=15,height=12, unit="cm",
     res=300, pointsize=12, compression="lzw", bg="white")
par(mar=c(4,5,3,5), las=1)

plot(x, y1, type="p", xaxt="n", col="red", pch=16,
     main="Two Axis Plot", xlab="x label",
     ylab="1st y axis label")
axis(1)
par(new=TRUE)
plot(x, y2, type="p", col="blue", pch=16,
     xaxt="n", yaxt="n", xlab="", ylab="")
axis(4)
par(las=0)
mtext("2nd y axis label", side=4, line=3, cex=1)
par(las=1)
legend("bottom", legend=c(legend1, legend2),
    lty=0, pch=c(16,16), col=c("red", "blue"),
    ncol=1, bty="o", cex=1,
    text.col="black", box.col="gray",
    inset=0.01)
dev.off()
```

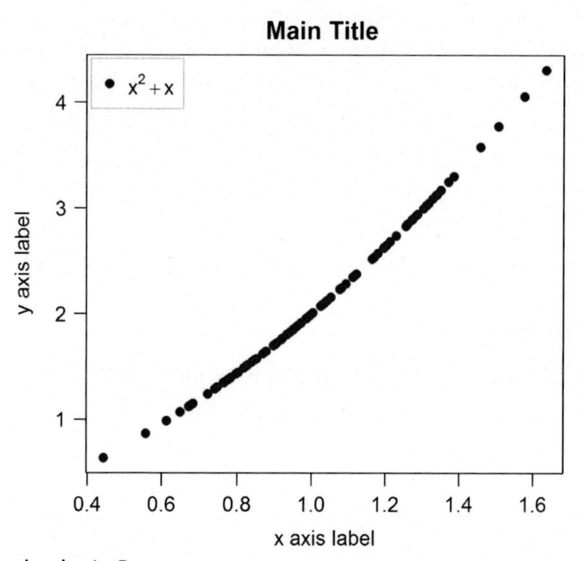

Fig. 3.1 A simple plot in R.

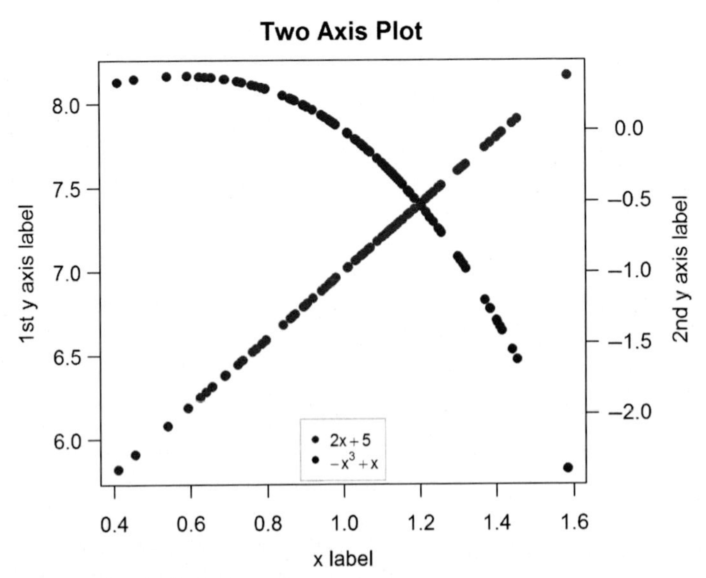

Fig. 3.2 A two-axis plot in R.

3.2 Plotting in Python

Matplotlib is a plotting library that produces high-quality graphics for publications and presentations. In matplotlib, users can save plots in various graphics formats, or it can be used interactively from IPython shell, jupyter notebook, and web interfaces of servers.

Matplotlib offers two programming interfaces to users. One of them is state machine oriented Pyplot API, which is easy to use with few lines of commands to generate basic plots like line plots, scatterplots, histograms, and bar plots. Learning pyplot is easy, and plotting commands in pyplot are very close to that of MATLAB. On the other hand, Matplotlib API has an object-oriented approach to deal with more complex and harder plotting tasks. It provides more commands for users to take control of plots, and it enables customization of plots according to the needs of users [2–4].

3.2.1 Pyplot API

Pyplot offers the easiest solution to produce plots. At the top of your plot commands, pyplot should be imported; usually it is also necessary to import numpy for numerical operations.

```
import matplotlib.pyplot as plt
import numpy as np
```

Following these lines, data should be created or imported from a file. To create a randomly distributed x variable with mean of m and standard deviation of sd and independent y as a function of x, you need to type these commands:

```
m, sd = 1.0, 0.25
x = np.random.normal(m, sd, 100)
y = x**2 + x
```

Prior to plot command, you need to define figure as an object by using plt.figure() command. Then, plot command is given to plot y versus x. Plot command accepts x,y variables in different formats; x,y could be numeric variables for 2D plotting, or time-series plots can be produced using a single variable with index values. Plot command can also be used for plotting multiple data series.

By default, line plots are generated for a given set of x,y variables. A scatter plot can be produced by defining marker type as an extra argument. In below example, marker type 'o' was used.

```
fig = plt.figure()

plt.plot(x, y, 'o', label='$y=x^2+x$')
```

After plot command, some extra commands can be used to add information to the plot, such as title, axis labels, and legend. At the end of these commands, plot can be visualized with plt.show() command (Fig. 3.3).

```
plt.xlabel('x axis label')
plt.ylabel('y axis label')

plt.title("Plot Title")

plt.legend()

plt.show()
```

3.2.2 Matplotlib API

Matplotlib API brings more customization for plots over pyplot. Most of the plot elements like axis, ticks, lines, and texts can be controlled by using functions in axes class. It is possible to draw many plots like line, scatter, step, log, vertical and horizontal bars, stem, pie, broken bar, filled polygon, spectral, box and whisker, violin, hexagonal binning, histogram, line contour, filled contour, image, and quiver.

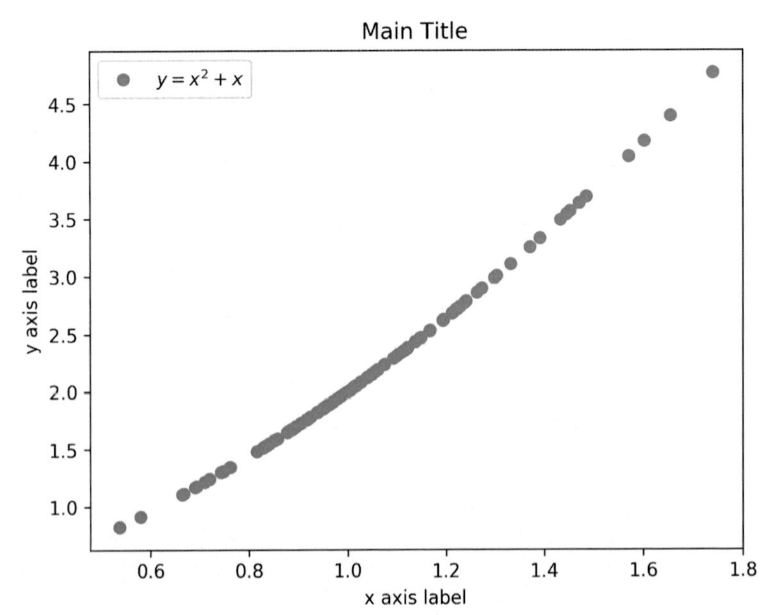

Fig. 3.3 A simple plot in Python.

An example plot produced by using Matplotlib API (Fig. 3.4):

```
import matplotlib.pyplot as plt
import numpy as np
from numpy import *

def two_axis(ax1, xdata, data1, data2, col1, col2):
    ax2 = ax1.twinx()
    ax1.plot(xdata, data1, '.', color=col1, label='$y1=2*x+5$')
    ax1.set_xlabel('x label')
    ax1.set_ylabel('1st y axis label')
    ax2.plot(xdata, data2, 'o', color=col2, label='$y2=-x^3+x$')
    ax2.set_ylabel('2nd y axis label')
    return ax1, ax2
```

Create data

```
m, sd = 1.0, 0.25
x = np.random.normal(m, sd, 100)
y1 = 2*x+5
y2 = -x**3+x
```

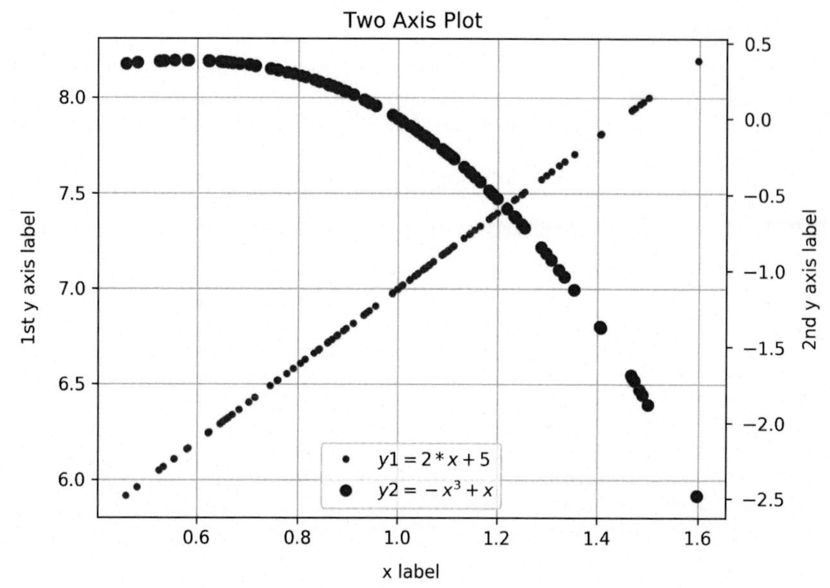

Fig. 3.4 A two-axis plot in Python.

Create plot and axes

```
fig, ax = plt.subplots()
ax1, ax2 = two_axis(ax, x, y1, y2, 'r', 'b')
```

Create legends

```
leg1, = ax1.plot(x, y1, '.', color='r')
leg2, = ax2.plot(x, y2, 'o', color='b')

plt.legend((leg1,leg2),('$y1=2*x+5$','$y2=-x^3+x$'),
loc='lower center')

plt.title('Two Axis Plot')

ax.grid(True)

plt.savefig('/path-to-figure-folder/figure-name.png',
format='png', dpi=300)

plt.show()
```

References

[1] R Manuals, An Introduction to R – Graphical procedures, https://cran.r-project.org/doc/manuals/r-release/R-intro.html#Graphics. Accessed 1 January 2018.
[2] Python Documentation, Matplotlib API, https://matplotlib.org/api/index.html. Accessed 1 January 2018.
[3] Python Documentation, Axes API, https://matplotlib.org/api/axes_api.html. Accessed 1 January 2018.
[4] Python Documentation, Pyplot API, https://matplotlib.org/api/pyplot_summary.html. Accessed 1 January 2018.

CHAPTER 4

Physical oceanography examples

4.1 Vertical profiling plots in R

The following dataset will be used to produce vertical profiling plots. There are 23 observations in the dataset, and vertical temperature and salinity data is given. Values do not belong to field observations.

```
      depth     temp     sal
1        10    12.71    35.41
2        50    12.25    35.32
3       100    11.08    35.30
4       250     8.46    34.95
5       500     6.46    34.72
6       750     4.08    34.53
7      1000     2.49    34.51
8      1250     2.35    34.56
9      1500     2.18    34.65
10     1750     1.95    34.74
11     2000     1.84    34.76
12     2250     1.72    34.78
13     2500     1.55    34.81
14     2750     1.43    34.79
15     3000     1.29    34.78
16     3250     1.09    34.77
17     3500     0.92    34.76
18     3750     0.79    34.75
19     4000     0.56    34.74
20     4250     0.35    34.72
21     4500     0.12    34.71
22     4750    -0.35    34.69
23     4900    -1.20    34.68
```

A vertical profiling plot for two variables in separate subplots (Fig. 4.1):

```
# Define working folder path and load data file
# Example path for Windows: "C:/Users/username/Documents/"
# Example path for Linux: "/home/username/Documents/"
# Example path for Mac OS: "/Users/username/Documents/"
```


https://doi.org/10.1016/B978-0-12-813491-7.00004-X

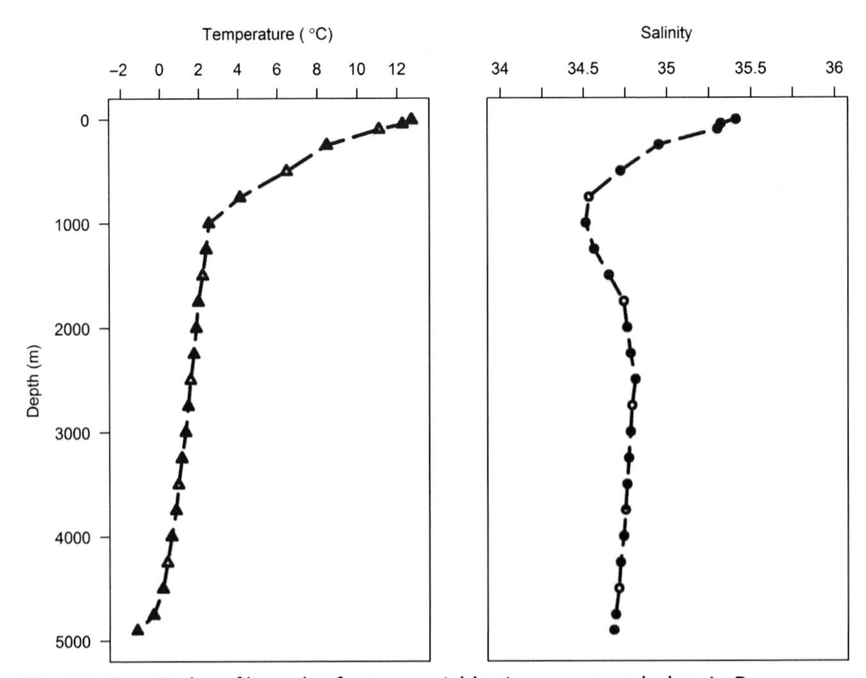

Fig. 4.1 A vertical profiling plot for two variables in separate subplots in R.

```
setwd("/path-to-local-data-folder/")
data <- read.table("data.txt", header=TRUE)
attach(data)
```

Find minimum and maximum values

```
mint <- floor(min(data['temp']))
maxt <- ceiling(max(data['temp']))
mins <- floor(min(data['sal']))
maxs <- ceiling(max(data['sal']))
```

Set axis ticks at defined intervals

```
t_ticks <- seq(mint,maxt,2)
s_ticks <- seq(mins,maxs,0.25)
y_ticks <- seq(0,5000,1000)
```

Define output folder path, figure size, and resolution

```
png("figure-name.png", width=19, height=14, unit="cm", res=600,
pointsize=10, bg="white")
```

Define locations of subplots with mfcol or mfrow parameter
Define subplot margins

```
par(mfcol=c(1,2), mar=c(1,5,4,1), las=1, xpd=TRUE)
```

Define the plot
Note that the axes and lines are not drawn with plot function; they are plotted with axis and lines functions manually
Axes are numbered as 1: bottom, 2: left, 3: top and 4: right.
To use top axis as x-axis, use 3 in axis function
To use left axis as y-axis, use 2 in axis function
Axis labels are drawn with mtext function; y axis label is rotated vertically using "las=0" in mtext function
Plot device is closed and saved with dev.off() function

```
plot(0,0, type="o", xaxt="n", yaxt="n", xlim=c(mint,maxt),
ylim=rev(range(0,5000)), col="white", pch=0, cex=1, lwd=2,
main="",  xlab="", ylab="")

axis(2,at=y_ticks,labels=y_ticks,cex.axis=1)

axis(3,at=t_ticks,labels=t_ticks,cex.axis=1)

mtext(expression("Temperature ("~degree*"C)"), side=3,
line=2.6, cex=1,  las=1, col="black")

mtext("Depth (m)", side=2, line=3.5, cex=1, las=0, col="black")

lines(data$temp,data$depth, col="blue", pch=2, type="o",
lty="longdash", lwd=3)

par(mar=c(1,2,4,2))

plot(0,0, type="o", xaxt="n", yaxt="n", xlim=c(mins,maxs),
ylim=rev(range(0,5000)), col="white", pch=0, cex=1, lwd=2,
main="",  xlab="", ylab="")

axis(3,at=s_ticks,labels=s_ticks,cex.axis=1)
```

Continued

```
mtext(expression("Salinity"), side=3, line=2.6, cex=1, las=1,
col="black")

lines(data$sal,data$depth, col="darkred", pch=1, type="o",
lty="longdash",  lwd=3)

dev.off()
```

4.2 Time-series plots in R

The following dataset will be used to produce time-series plots. There are 64 observations in the dataset. There are three variables: date, density, and salinity. Values do not belong to field observations.

	date	dens	salin
1	2003-09-19	25.92	36.42
2	2003-09-18	28.82	36.90
3	2003-09-17	29.43	38.33
4	2003-09-16	25.84	36.81
5	2003-09-15	28.08	39.51
6	2003-09-12	29.25	37.64
7	2003-09-11	27.93	35.39
8	2003-09-10	29.69	38.21
9	2003-09-09	26.26	38.31
10	2003-09-08	29.69	38.78
11	2003-09-05	29.22	38.70
12	2003-09-04	29.74	35.16
13	2003-09-03	29.94	34.85
14	2003-09-02	25.22	37.60
15	2003-08-29	27.62	37.38
16	2003-08-28	26.22	37.58
17	2003-08-27	26.47	38.25
18	2003-08-26	28.93	38.42
19	2003-08-25	28.62	37.44
20	2003-08-22	29.86	36.99
21	2003-08-21	28.32	39.96
22	2003-08-20	29.48	37.57
23	2003-08-19	24.96	35.11
24	2003-08-18	27.40	37.15
25	2003-08-15	26.86	37.24
26	2003-08-14	27.95	39.49

```
27        2003-08-13        25.38        37.96
28        2003-08-12        24.54        38.28
29        2003-08-11        25.04        34.43
30        2003-08-08        25.51        37.52
31        2003-08-07        25.78        36.26
32        2003-08-06        28.88        35.42
33        2003-08-05        27.34        39.68
34        2003-08-04        26.46        37.10
35        2003-08-01        27.49        37.71
36        2003-07-31        28.82        39.25
37        2003-07-30        28.73        38.08
38        2003-07-29        25.70        37.76
39        2003-07-28        26.76        35.85
40        2003-07-25        26.95        37.15
41        2003-07-24        29.86        33.80
42        2003-07-23        26.91        38.15
43        2003-07-22        27.59        36.38
44        2003-07-21        29.76        37.29
45        2003-07-18        28.02        36.65
46        2003-07-17        27.81        38.41
47        2003-07-16        29.36        37.10
48        2003-07-15        29.54        39.36
49        2003-07-14        28.42        37.93
50        2003-07-11        27.78        36.45
51        2003-07-10        26.94        40.15
52        2003-07-09        25.89        36.15
53        2003-07-08        29.25        37.37
54        2003-07-07        27.99        39.50
55        2003-07-03        26.27        34.05
56        2003-07-02        27.22        36.89
57        2003-07-01        28.20        36.32
58        2003-06-30        28.77        39.11
59        2003-06-27        28.67        35.73
60        2003-06-26        25.80        39.06
61        2003-06-25        28.97        35.36
62        2003-06-24        24.92        41.03
63        2003-06-23        25.75        36.35
64        2003-06-20        25.48        37.13
```

A time-series plot for a single variable (Fig. 4.2):

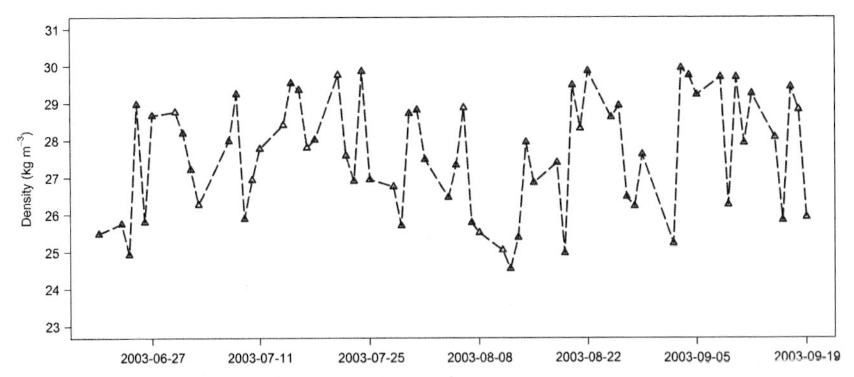

Fig. 4.2 A time-series plot for a single variable in R.

Define working folder path and load data file

```
setwd("/path-to-local-data-folder/")
```

Introduce date values in ISO format

```
data$date = as.Date(data$date, format='%Y-%m-%d')
```

Define axis tick labels manually with seq function

```
y1list <- seq(floor(min(data$dens))-1,
ceiling(max(data$dens))+1, 1)

y2list <- seq(floor(min(data$salin)),
ceiling(max(data$salin)), 1)
```

Define x axis tick values

```
xaxpos <- c(data$date[6], data$date[15], data$date[25],
data$date[35], data$date[45], data$date[54], data$date[64])
```

Define output folder path for figure and set figure parameters

```
png("figure-name.png", width=28, height=12, unit="cm", res=600,
pointsize=10, bg="white")

par(mar=c(2,4.5,2,2), las=1, xpd=TRUE)
```

Define the plot
Define and format x axis with axis.Date() function

```
plot(data$date,data$dens,type="o", xaxt="n", yaxt="n",
axes=FALSE,  col="white", pch=0, cex=1, lwd=2, main="",
xlab="", ylab="",  cex.axis=1.2, cex.lab=1.2,
ylim=c(floor(min(data$dens))-1, ceiling(max(data$dens))+1))

axis.Date(1, at=xaxpos, format="%Y-%m-%d", las=1)

axis(2,at=yllist,labels=yllist,cex.axis=1)

lines(data$date,data$dens, col="blue", pch=2, type="o",
lty="longdash", lwd=2)

par(las=0)

mtext(2,text=expression("Density (kg m"o-3*")"),line=2.5,cex=1)

par(las=1)

box()

dev.off()
```

4.3 Temperature-salinity diagrams in R

In this example, data given in vertical profiling plots at section 1 is used.

A temperature-salinity diagram includes contours of density as a function of temperature and salinity. For calculation of seawater density, sw_dens function is used from marelac package. In this example, data are plotted as scatter points and colored according to their values. For contour plot, contour2D function from plot3D package is used.

This example (Fig. 4.3) is a modified version of the original source code implemented by Karline Soetaert at R for Science website (http://www.rforscience.com/portfolio/salinity/) [1].

Define working folder path and load data file
Load libraries: marelac and plot3D

```
setwd("/path-to-local-data-folder/")
library(marelac)
library(plot3D)
data <- read.table("data.txt", header=TRUE, sep="\t")
```

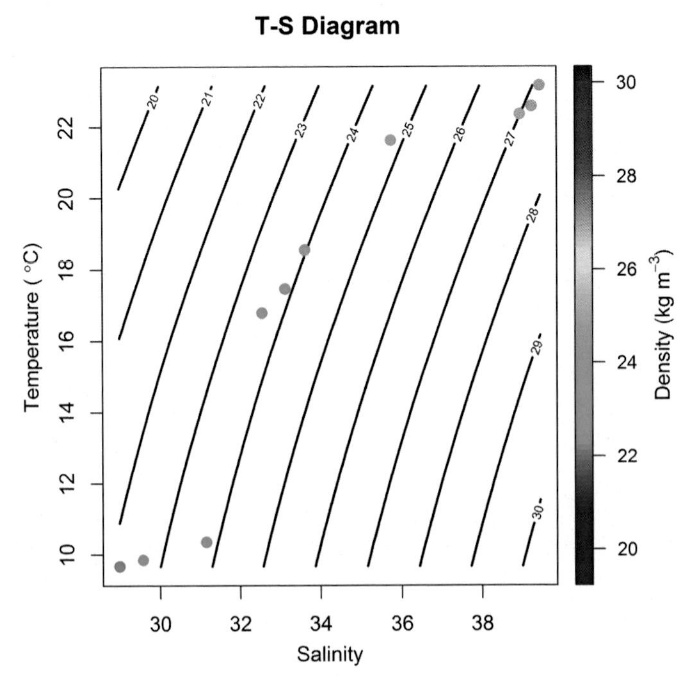

Fig. 4.3 A temperature-salinity diagram in R.

Find minimum and maximum values for data

```
mint = min(data['temp'])
maxt = max(data['temp'])
mins = min(data['sal'])
maxs = max(data['sal'])
```

Calculate seawater density from temperature and salinity data
Generate density values for plotting contour lines between min-max temperature and salinity data

```
sal.c <- seq(from=mins, to=maxs, length.out=100)

temp.c <- seq(from=mint, to=maxt, length.out=100)

sigmat.c <- outer(sal.c, temp.c, FUN=function(S, t)
sw_dens(S=S, t=t) - 1000)
```

Define output folder path for figure and set figure parameters

```
png(filename="figure-name.png", width=15, height=15, unit="cm",
res=600, pointsize=12, bg="white")

par(mar=c(5,5,4,6))
```

Plot contours of density

```
contour2D(x=sal.c, y=temp.c, z=sigmat.c, lwd=2, main="T-S
Diagram", col="black",xlab=expression("Salinity"),
ylab=expression("Temperature ("~degree*"C)"))
```

Prepare real data for scatter plot

```
temp <- unlist(data['temp'], use.names=FALSE)
sal <- unlist(data['sal'], use.names=FALSE)
sigma_theta <- sw_dens(S = sal, t = temp) - 1000
```

Plot scatter points of real data
Colors and colorbar ranges given according to density values

```
scatter2D(sal, temp, colvar=sigma_theta, pch=16, cex=1.25,
add=TRUE, clim=range(sigmat.c), colkey=FALSE)

colkey(clim=range(sigmat.c), dist=0.005, side=4, add=TRUE,
clab=expression("Density (kg m"^-3*")"), col.clab="black",
side.clab=4, line.clab=2.5, length=1, width=0.8,
col.axis="black",  col.ticks="black", cex.axis=0.9)

dev.off()
```

4.4 Maps in R

In R, it is possible to produce maps with ready-to-use data sets in maps, mapdata, rworldmap, rworldxtra, PBSmapping, and oceanmap packages. Maps like GSHHG coastline can also be plotted by using maptools package. It is also possible to get maps from Google Maps, OpenStreetMap, Stamen Maps, or Naver Map servers with ggmap package. OpenStreetMap package is an another way to get maps from OpenStreetMap servers.

GSHHG coastline data used in this example is provided, developed, and maintained by Paul Wessel from SOEST, University of Hawai'i, Honolulu, USA, and Walter H. F. Smith from NOAA Geosciences Lab, National Ocean Service, Silver Spring, USA at https://www.soest.hawaii.edu/pwessel/gshhg/ website [2].

A simple map with text, points, north arrow, and map scale (Fig. 4.4):

Coordinate data including lower left and upper right limits, sampling points, and texts:

```
name                 lon     lat
lower-left-point     -9.87   28.33
upper-right-point    51.92   47.85
"Mediterranean Sea"  13.15   33.60
"Black Sea"          30.65   42.90
1                    29.41   33.81
2                    25.84   33.88
3                    20.76   35.54
4                    16.85   36.40
5                    12.85   35.83
```

Define working folder path and load coordinate data file

```
setwd("/path-to-local-data-folder/")
library(maptools)
library(PBSmapping)
library(GISTools)
library(geosphere)
dataset <- read.table("data.txt", header=TRUE)
```

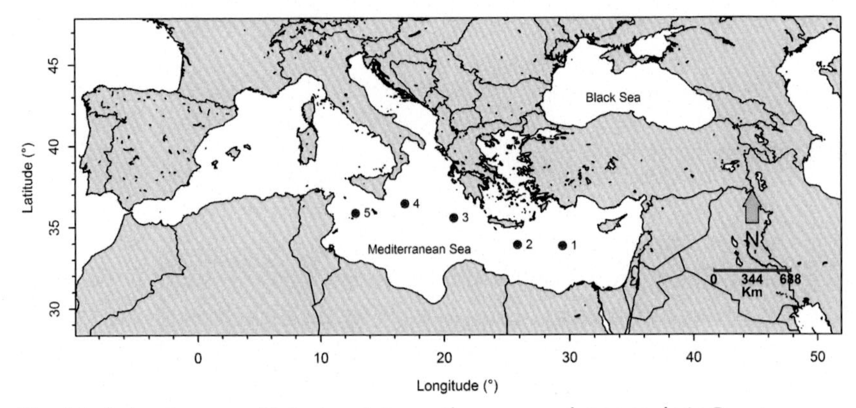

Fig. 4.4 A simple map with text, points, north arrow, and map scale in R.

Read data columns into vector variables

```
lon <- dataset$lon
lat <- dataset$lat
label <- dataset$name
```

Define min-max lat-lon limits

```
xlim <- c(lon[1], lon[2])
ylim <- c(lat[1], lat[2])
```

Define positions of map scale and north arrow
Distance between two points is calculated with distGeo function from geosphere package
distGeo calculates the distance using geodesic ellipsoid

```
sblon1 <- lon[2]-((lon[2]-lon[1])/10)-lon[2]*0.02
sblon2 <- lon[2]-lon[2]*0.02
sblat <- lat[1]+lat[1]*0.1

p1 <- c(sblat, sblon1)
p2 <- c(sblat, sblon2)

dist <- distGeo(p1, p2)/1000
d1 <- dist/2
```

Import GSHHG coastline and border data

```
coasts <- importGSHHS("/path-to-data-folder/gshhs_f.b",
xlim=xlim, ylim=ylim, maxLevel=4)

borders <- importGSHHS("/path-to-data-folder/wdb_borders_f.b",
xlim=xlim, ylim=ylim, maxLevel=4)
```

Define output folder path for figure and set figure parameters

```
png(filename="figure-name.png", width=19, height=14, unit="cm",
res=600, pointsize=10, bg="white")

par(oma=c(0,0,0,0),mar=c(0,0,0,0),xpd=TRUE)
```

Define the plot
Plot basemap with plotMap function and add borders with addLines function from PBSmapping package
Plot points with points function
Plot text with text function
Plot map scale with map.scale function from GISTools package
Plot north arrow with north.arrow function from GISTools package

```
plotMap (coasts, tck=c(-0.02), plt=c(0.16, 0.97, 0.16, 0.97),
col="gray90")

addLines (borders)

points(lon[5:9], lat[5:9], col="black", bg="red", pch=21)

text(lon[3:4], lat[3:4], labels=label[3:4], cex=0.7, pos=4)

text(lon[5:9], lat[5:9], labels=label[5:9], cex=0.7, pos=4)

north.arrow(xb=sblon1, yb=sblat+4, len=0.5, lab="N",
col="gold", cex.lab=1.2)

par(cex=0.75, font=2)

map.scale(xc=sblon1, yc=sblat+1, len=(sblon2-sblon1),
units="Km", ndivs=2, subdiv=round(dl), tcol='black',
scol='darkblue', sfcol='gold')

dev.off()
```

4.5 Transect plots in R

Transect plots can be generated using image2D function from plot3D package. In this example, World Ocean Atlas (WOA) 2013v2 Annual Mean Temperature data [3] is used. WOA13v2 data is provided by the NOAA/NCEI OCL, USA, from their website at https://www.nodc.noaa.gov/OC5/woa13/ [4].

WOA13v2 1° grid annual temperature data can be downloaded from:

https://data.nodc.noaa.gov/thredds/fileServer/woa/WOA13/DATAv2/temperature/netcdf/decav/1.00/woa13_decav_t00_01v2.nc

This example is a modified version of the original source code implemented by Karline Soetaert at R for Science website (http://www.rforscience.com/portfolio/silicate-phosphate-and-nitrate/) [5].

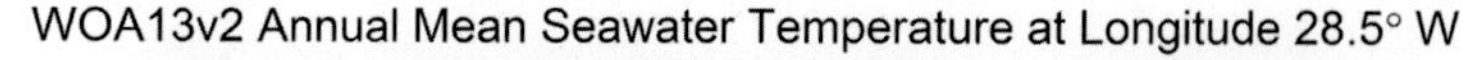

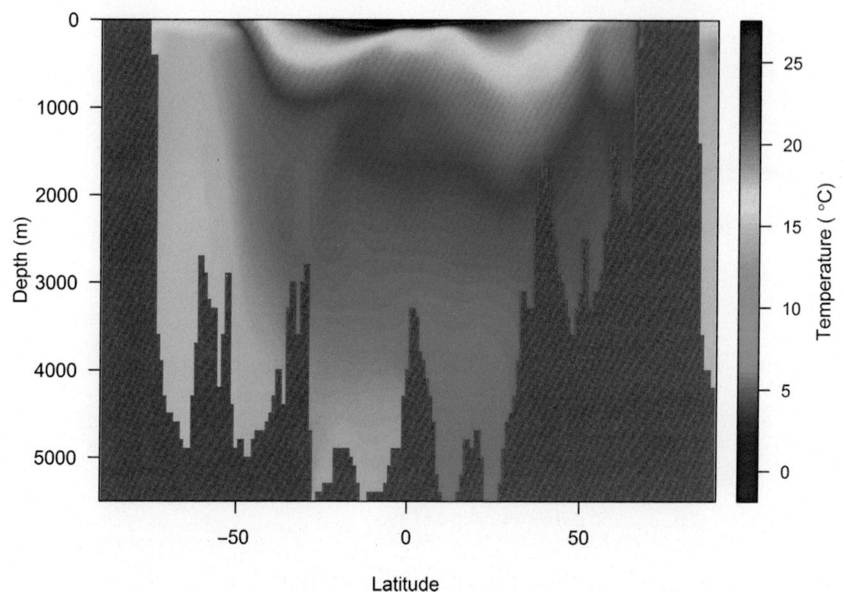

Fig. 4.5 A transect plot along longitude 28.5 °W in R.

A transect plot along longitude 28.5 °W (Fig. 4.5):

Define working folder path and load data file

```
setwd("/path-to-local-data-folder/")
library(RNetCDF)
library(oceanmap)
library(plot3D)
data(cmap)
```

Load NetCDF data file

```
path <- "/path-to-file/woa13_decav_t00_01v2.nc"
```

Read data file into variables and create a list of variables

```
D.nc <- open.nc(paste(path, sep=""))
lon <- var.get.nc(D.nc, 'lon')
lat <- var.get.nc(D.nc, 'lat')
time <- var.get.nc(D.nc, 'time')
depth <- var.get.nc(D.nc, 'depth')
```

Continued

```
value <- var.get.nc(D.nc, 't_an')
name <- att.get.nc(D.nc, 't_an', "long_name")
units <- att.get.nc(D.nc, 't_an', "units")
lon <- c(rev(lon[1:180])*-1, lon[181:360]+180)
temp <- list(lon=lon, lat=lat, depth=depth, time=time,
value=value, name=name, units=units)
```

Define longitude (180-28.5) to extract only transect data

```
long <- 151.5
```

Define output folder path for figure and set figure parameters

```
png("figure-name.png", width=16, height=12, unit="cm",
res=600,  pointsize=10, bg="white")

par(mar=c(5,5,3,6), las=1, xpd=TRUE)
```

Define the transect plot with image2D function
Color key is defined separately with colkey function

```
image2D(z=temp$value[long, , ], x=temp$lat, y=temp$depth,
colkey=FALSE,  ylim=c(5500, 0), xlim=c(-90,90), yaxp=c(-
90,90,6), col=cmap$sst,  NAcol="gray30", resfac=4,
xlab="Latitude", ylab="Depth (m)", main=expression("Transect
plot for WOA13v2 Annual Mean SST at Longitude
28.5"*degree~"W"))

par(las=0)

colkey(clim=range(temp$value[long, , ], na.rm=T),
dist=0.005, side=4, add=TRUE, clab=expression("Temperature
("~degree*"C)"), col.clab="black", side.clab=4,
line.clab=2.5, length=1, width=0.8, col.axis="black",
col.ticks="black", cex.axis=0.9)

dev.off()
```

4.6 Surface plots in R

Surface plots can also be generated using image2D function from plot3D package. In this example, World Ocean Atlas (WOA) 2013v2 Annual

Mean Temperature data [3] is used. WOA13v2 data is provided by the NOAA/NCEI OCL, USA, from their website at https://www.nodc.noaa.gov/OC5/woa13/ [4].

WOA13v2 1° grid annual temperature data can be downloaded from:

https://data.nodc.noaa.gov/thredds/fileServer/woa/WOA13/DATAv2/temperature/netcdf/decav/1.00/woa13_decav_t00_01v2.nc

This example is a modified version of the original source code implemented by Karline Soetaert at R for Science website (http://www.rforscience.com/portfolio/silicate-phosphate-and-nitrate/) [5].

A surface plot for worldwide annual mean SST data from WOA13v2 database (Fig. 4.6):

Define working folder path and load data file

```
setwd("/path-to-local-data-folder/")
library(plot3D)
library(RNetCDF)
library(maps)
```

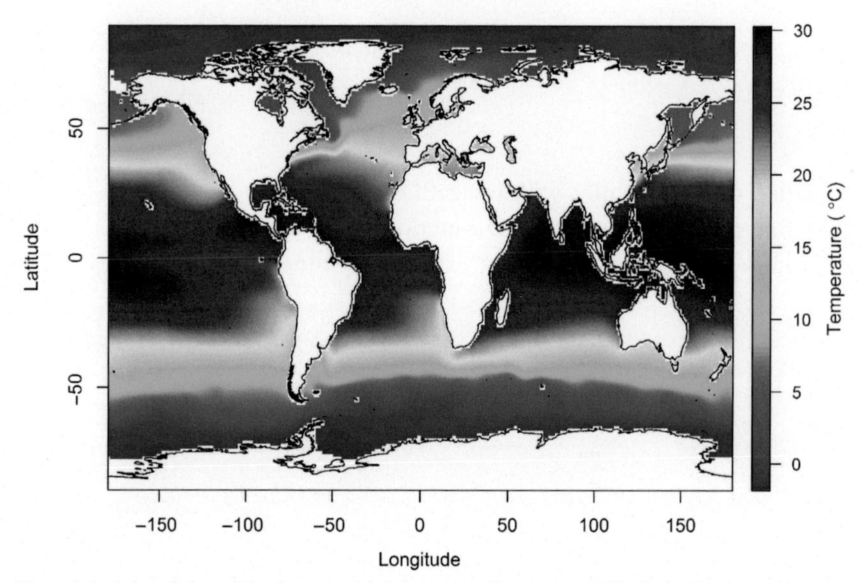

Fig. 4.6 A surface plot for worldwide annual mean SST data from WOA13v2 database in R.

Define coastline data from "world" map in maps package

```
m <- map("world", plot=FALSE, interior=FALSE)
```

Load NetCDF data file

```
path <- "/path-to-file/woa13_decav_t00_01v2.nc"
```

Read data file into variables and create a list of variables

```
D.nc <- open.nc(paste(path, sep=""))
lon <- var.get.nc(D.nc, 'lon')
lat <- var.get.nc(D.nc, 'lat')
time <- var.get.nc(D.nc, 'time')
depth <- var.get.nc(D.nc, 'depth')
value <- var.get.nc(D.nc, 't_an')
name <- att.get.nc(D.nc, 't_an', "long_name")
units <- att.get.nc(D.nc, 't_an', "units")
temp <- list(lon=lon, lat=lat, depth=depth, time=time,
value=value, name=name, units=units)
```

Define output folder path for figure and set figure parameters

```
png(filename="sst_surface_r.png", width=20, height=15,
unit="cm", res=600, pointsize=12, bg="white")

par(oma = c(0, 1, 0, 4))
```

Define the surface plot with image2D function
Define coastlines with lines2D function
Color key is defined separately with colkey function

```
image2D(z=temp$value[ , ,1], x=temp$lon, y=temp$lat, resfac=4,
colkey=FALSE, main="WOA13v2 Annual Mean SST",
xlab="Longitude", ylab="Latitude", clab="")

lines2D(x=m$x, y=m$y, add=TRUE, col="black")

colkey(clim=range(temp$value[ , ,1], na.rm=T), dist=0.005,
side=4, add=TRUE, clab=expression("Temperature
("~degree*"C)"), col.clab="black", side.clab=4,
line.clab=2.5, length=1, width=0.8, col.axis="black",
col.ticks="black", cex.axis=0.9)

dev.off()
```

4.7 Vertical profiling plots in Python

In this example, data given in vertical profiling plots at section 1 is used.

A vertical profiling plot for two variables in separate subplots (Fig. 4.7):
Load required libraries

```
import numpy as np
import matplotlib.pyplot as plt
```

Load data file

```
f = open('/ path-to-local-data-folder/data.txt', 'r')
data = np.genfromtxt(f, dtype="float", delimiter='\t',
names=True)
f.close()
```

Define variables from data columns

```
depth = data['depth']
temp = data['temp']
sal = data['sal']
del(data)
```

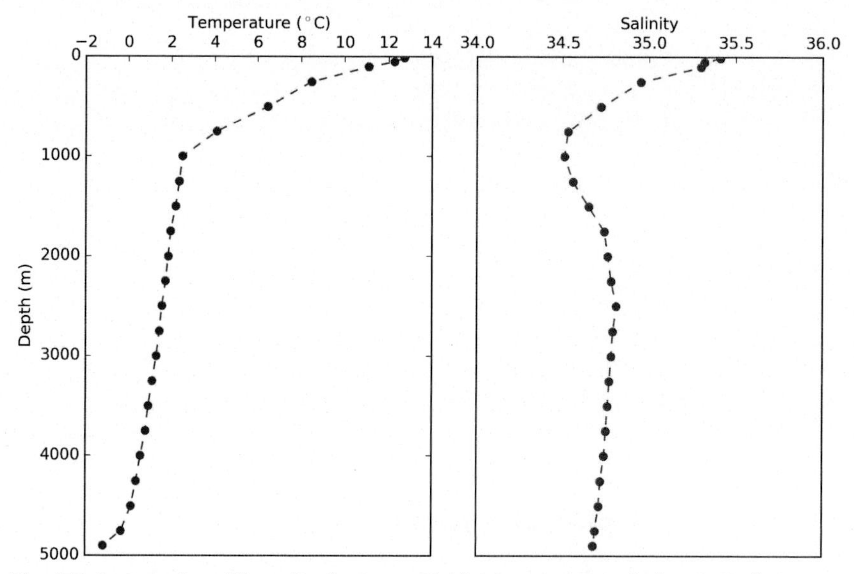

Fig. 4.7 A vertical profiling plot for two variables in separate subplots in Python.

Define intervals of axis ticks

```
mins = np.floor(np.min(sal))
maxs = np.ceil(np.max(sal))
```

Plot data
Define subplot 1

```
fig1, (ax1, ax2) = plt.subplots(1,2,sharey=True,figsize=(9, 6))

ax1.plot(temp,depth,'o--b')

ax1.set_ylim(ax1.get_ylim()[::-1]) # reverse y axis and set limits

ax1.set_xlabel('Temperature [$^\circ$C]') # x axis label

ax1.set_ylabel('Depth (m)') # y axis label

ax1.xaxis.set_label_position('top') # move x axis label to top

ax1.xaxis.set_ticks_position('top') # move x axis ticks to top
```

Define subplot 2

```
ax2.plot(sal,depth,'o--r')
ax2.set_xlabel('Salinity')
ax2.set_xlim(mins,maxs) # set x axis limits
ax2.xaxis.set_label_position('top')
ax2.xaxis.set_ticks_position('top')
ax2.yaxis.set_visible(False) # hide y axis for 2nd subplot
```

Set plot margins to tight layout; save and show figure

```
plt.tight_layout()

plt.savefig('/path-to-figure-folder/figure-name.png',
format='png', dpi=600, transparent=False)

plt.show()
```

4.8 Time series plots in Python

In this example, data given in time series plots in R at section 2 is used.

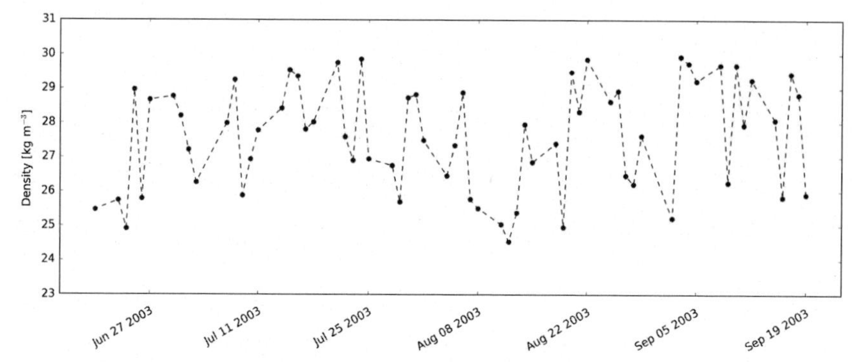

Fig. 4.8 A time-series plot for a single variable in Python.

A time-series plot for a single variable (Fig. 4.8):
Load required libraries

```
import datetime as dt
import numpy as np
import matplotlib.pyplot as plt
```

Define variables and load data

```
time = []
dens = []
dtime = []

time = np.loadtxt('/path-to-local-data-folder/date.txt',
delimiter='\t', skiprows=1, usecols=[0], dtype='str')

dtime = [dt.datetime.strptime(d,'%Y-%m-%d') for d in time]

dens = np.loadtxt('/path-to-local-data-folder/date.txt',
delimiter='\t', skiprows=1, usecols=[1], dtype='float')
```

Define intervals of axis ticks

```
ymin = np.floor(np.min(dens)-1)
ymax = np.ceil(np.max(dens)+1)
```

Plot data

```
fig = plt.figure(figsize=(15,6))

ax = plt.gca()
```

Continued

```
plt.plot_date(x=dtime, y=dens, marker='o', linestyle='dashed')

plt.xlabel('')

plt.ylabel('Density [kg m$^{-3}$]')

fig.autofmt_xdate()

ax.set_ylim([ymin,ymax])

plt.margins(0.05, 0.1)
```

Set plot margins to tight layout; save and show figure

```
plt.tight_layout()

plt.savefig('/path-to-figure-folder/figure-name.png',
format='png', dpi=600, transparent=False)

plt.show()
```

4.9 Temperature-salinity diagrams in Python

In this example, data given in vertical profiling plots at section 1 is used.

This example is a modified version of the original source code implemented by Juliana Leonel at "Figures for lectures in oceanography" website (https://juoceano.github.io/lecture_figures/) [6]. The original source code is licensed under a Creative Commons Attribution-ShareAlike 4.0 International License.

A temperature-salinity diagram (Fig. 4.9):
Load required libraries

```
import gsw
import numpy as np
import pandas as pd
import matplotlib.pyplot as plt
from matplotlib.ticker import MaxNLocator
```

Load data and sort values according to temperature column in ascending order

```
cd = pd.read_csv('/path-to-local-data-folder/data.txt',
sep='\t', header=0)

df = cd.sort_values(by='temp', ascending=True)
```

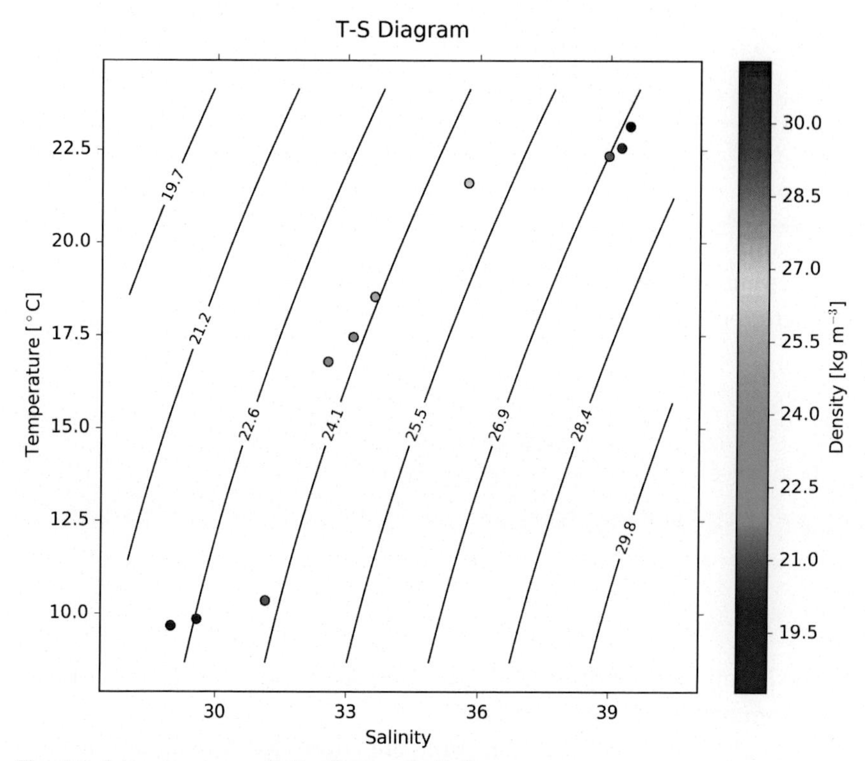

Fig. 4.9 A temperature-salinity diagram in Python.

Define min-max values of temperature and salinity

```
mint = np.min(df['temp'])
maxt = np.max(df['temp'])
mins = np.min(df['sal'])
maxs = np.max(df['sal'])
```

Generate linearly spaced input data for contour levels from min-max temperature and salinity data

```
temp = np.linspace(mint-1, maxt+1, 25)
sal = np.linspace(mins-1, maxs+1, 25)
```

Grid input data and calculate seawater density from temperature and salinity data

```
temp, sal = np.meshgrid(temp, sal)

sigma_theta = gsw.sigma0(sal, temp)

levels = np.linspace(sigma_theta.min(), sigma_theta.max(), 10)
```

Plot data
Contours are plotted with ax.contour function
Contour labels are plotted with plt.clabel function
Scatter points are plotted with plt.scatter function

```
fig, ax = plt.subplots(figsize=(8.25, 7))

cs = ax.contour(sal, temp, sigma_theta, colors='black',
levels=levels, zorder=1, extend='both')

cl = plt.clabel(cs, fontsize=10, inline=1, fmt = '%.1f')

sc = plt.scatter(df['sal'], df['temp'], c=levels, s=35)

cb = plt.colorbar(sc)
```

Maximum number of axis ticks are controlled with MaxNLocator function from matplotlib.ticker library
Plot and colorbar axis ticks directions are controlled with ax.tick_params and cb.ax.tick_params functions

```
ax.set_xlabel('Salinity')
ax.set_ylabel('Temperature [$^\circ$C]')
ax.set_title('T-S Diagram', y=1.025)
ax.xaxis.set_major_locator(MaxNLocator(nbins=6))
ax.yaxis.set_major_locator(MaxNLocator(nbins=8))
ax.tick_params(direction='out')
cb.ax.tick_params(direction='out')
cb.set_label('Density [kg m$^{-3}$]')
```

Set plot margins to tight layout; save and show figure

```
plt.tight_layout()

plt.savefig('/path-to-figure-folder/figure-name.png',
format='png', dpi=600, transparent=False)

plt.show()
```

4.10 Maps in Python

In this example, data given at Section 4.4 is used.

A simple map with text, points, north arrow, and map scale (Fig. 4.10):
Load required libraries

```
from __future__ import unicode_literals, division
from mpl_toolkits.basemap import Basemap
import numpy as np
import matplotlib.pyplot as plt
import matplotlib.ticker as ticker
import pandas as pd
from geopy.distance import distance
```

Load data and sort values according to temperature column in ascending order

```
cd = pd.read_csv('/path-to-local-data-folder/data.txt',
sep='\t', header=0)
```

Define map limits

```
lllon = cd['lon'][0]
lllat = cd['lat'][0]
urlon = cd['lon'][1]
urlat = cd['lat'][1]
```

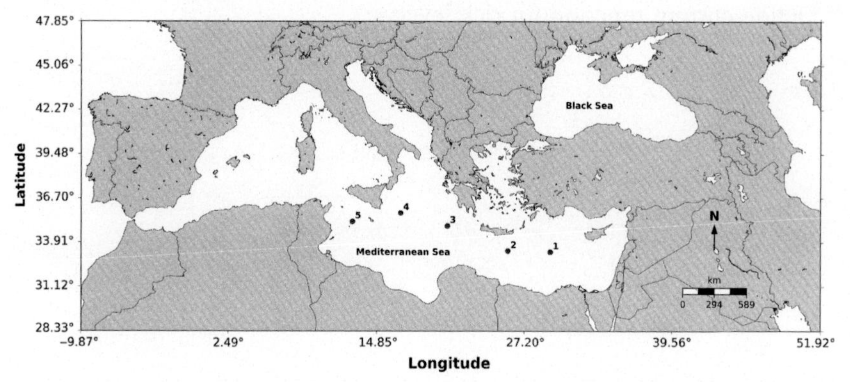

Fig. 4.10 A simple map with text, points, north arrow, and map scale in Python.

Create figure object and define figure size

```
fig = plt.figure(figsize=(16, 7))
ax = plt.axes()
```

Plot map with Basemap function from Basemap toolkit of matplotlib

```
m = Basemap(llcrnrlon=lllon, llcrnrlat=lllat, urcrnrlon=urlon,
urcrnrlat=urlat, resolution='h', projection='merc',
suppress_ticks=False)
```

Draw coastlines, rivers and countries; fill continents and lakes

```
m.drawcoastlines(color='black', linewidth=0.5)

m.drawmapboundary(fill_color='white')

m.fillcontinents(color='lightgrey', lake_color='white')

# m.drawrivers(linewidth=2, linestyle='solid',
color='darkblue', antialiased=1, ax=None, zorder=None)

m.drawcountries(linewidth=0.5,  linestyle='solid',  color='k',
antialiased=1, ax=None, zorder=None)
```

Define custom axis tick locations and labels
Number of ticks at x and y axes

```
numx_ticks = 6
numy_ticks = 8
```

Define custom function for tick levels

```
def ranges(min,max,n):
    step = (max-min)/n
    return ['{}'.format(u'%.2f\N{DEGREE SIGN}') % (min+step*i)
for i in range(n+1)]

xrange = ranges(lllon,urlon,numx_ticks-1)
yrange = ranges(lllat,urlat,numy_ticks-1)
```

Set tick locations and labels

```
ax.xaxis.set_major_locator(ticker.LinearLocator(numx_ticks))
ax.xaxis.set_major_formatter(ticker.FixedFormatter(xrange))

ax.yaxis.set_major_locator(ticker.LinearLocator(numy_ticks))
ax.yaxis.set_major_formatter(ticker.FixedFormatter(yrange))
```

Set tick positions to outer direction

```
ax.tick_params(direction='out')
```

Plot the stations

```
fs=12; fw='bold'; # plt.text arguments

s=40; mar='o'; fc='red'; ec='black'; # m.scatter arguments

xoff=0.2; yoff=0.2; # offset values for labels
```

Station 1

```
m.scatter(cd['lon'][4],cd['lat'][4],latlon=True,s=s, marker=mar,
facecolor=fc,edgecolor=ec)
lon = cd['lon'][4]+xoff
lat = cd['lat'][4]+yoff
x, y = m(lon, lat)
plt.text(x,y,cd['name'][4],fontsize=fs,fontweight=fw)
```

Station 2

```
m.scatter(cd['lon'][5],cd['lat'][5],latlon=True,s=s, marker=mar,
facecolor=fc,edgecolor=ec)
lon = cd['lon'][5]+xoff
lat = cd['lat'][5]+yoff
x, y = m(lon, lat)
plt.text(x,y,cd['name'][5],fontsize=fs,fontweight=fw)
```

Station 3

```
m.scatter(cd['lon'][6],cd['lat'][6],latlon=True,s=s,marker=mar,
facecolor=fc,edgecolor=ec)
lon = cd['lon'][6]+xoff
lat = cd['lat'][6]+yoff
x, y = m(lon, lat)
plt.text(x,y,cd['name'][6],fontsize=fs,fontweight=fw)
```

Station 4

```
m.scatter(cd['lon'][7],cd['lat'][7],latlon=True,s=s,marker=mar,
facecolor=fc,edgecolor=ec)
lon = cd['lon'][7]+xoff
lat = cd['lat'][7]+yoff
x, y = m(lon, lat)
plt.text(x,y,cd['name'][7],fontsize=fs,fontweight=fw)
```

Station 5

```
m.scatter(cd['lon'][8],cd['lat'][8],latlon=True,s=s,marker=mar,
facecolor=fc,edgecolor=ec)
lon = cd['lon'][8]+xoff
lat = cd['lat'][8]+yoff
x, y = m(lon, lat)
plt.text(x,y,cd['name'][8],fontsize=fs,fontweight=fw)
```

Plot the names of cities, seas, and countries
Mediterranean Sea

```
lon = cd['lon'][2]
lat = cd['lat'][2]
v, z = m(lon, lat)
plt.text(v,z,cd['name'][2],fontsize=fs,fontweight=fw)
```

Black Sea

```
lon = cd['lon'][3]
lat = cd['lat'][3]
v, z = m(lon, lat)
plt.text(v,z,cd['name'][3],fontsize=fs,fontweight=fw)
```

Add scale bar with m.drawmapscale function
Location of the scale bar on plot is automatically determined with sblon and sblat
Distance between min-max scalebar points are calculated with distance().km function from geopy library
Calculated distance is used in drawmapscale function with length parameter

```
sblon1 = urlon-((urlon-lllon)/10)-urlon*0.05
sblon2 = urlon-urlon*0.05
sblat = lllat+lllat*0.1

p1 = [sblat, sblon1]
p2 = [sblat, sblon2]

dist = distance(p1, p2).km

m.drawmapscale(sblon1, sblat, sblon2, sblat,
    length=dist,
    units='km', fontsize=fs,
    yoffset=None,
    barstyle='fancy', labelstyle='simple',
    fillcolor1='w', fillcolor2='#000000',
    fontcolor='#000000',
    zorder=1)
```

Add north arrow with plt.arrow function
Location of north arrow on the plot is automatically determined with sblon and sblat

```
offset = sblon1*0.01
lon1 = sblon1
lat1 = sblat + (sblat-lllat)
lat2 = sblat + ((urlat-lllat)/12) + (sblat-lllat)
lat3 = lat2 + (lat2-lat1)/5
v1, z1 = m(lon1,lat1)
v2, z2 = m(lon1,lat2)
v3, z3 = m(lon1-offset,lat3)
plt.arrow(v1, z1, v2-v1, z2-z1, fc='black', ec='black',
width=2500, length_includes_head=True)
plt.text(v3, z3, 'N', fontweight='bold', fontsize=17)
```

Add plot title, set plot margins to tight layout, save and show figure

```
plt.title('')

plt.ylabel('Latitude', labelpad=30, fontsize=15,
fontweight='bold')

plt.xlabel('Longitude', labelpad=20, fontsize=15,
fontweight='bold')

plt.tight_layout()

plt.savefig('/path-to-figure-folder/figure-name.png',
format='png', dpi=600, transparent=False)

#plt.show()
```

4.11 Transect plots in Python

Transects can be plotted in Python using plot_section function from ctd library. In this example, World Ocean Atlas (WOA) 2013v2 Annual Mean Temperature data [3] is used. WOA13v2 data is provided by the NOAA/NCEI OCL, USA, from their website at https://www.nodc.noaa.gov/OC5/woa13/ [4].

WOA13v2 1° grid annual temperature data can be downloaded from: https://data.nodc.noaa.gov/thredds/fileServer/woa/WOA13/DATAv2/temperature/netcdf/decav/1.00/woa13_decav_t00_01v2.nc

This example is a modified version of the original source code implemented by Juliana Leonel at "Figures for lectures in oceanography" website (https://juoceano.github.io/lecture_figures/) [6]. The original source code is licensed under a Creative Commons Attribution-ShareAlike 4.0 International License.

A transect plot along longitude 28.5 °W (Fig. 4.11):

Load required libraries

```
import warnings
warnings.filterwarnings('ignore')

import iris
iris.FUTURE.netcdf_promote = True
```

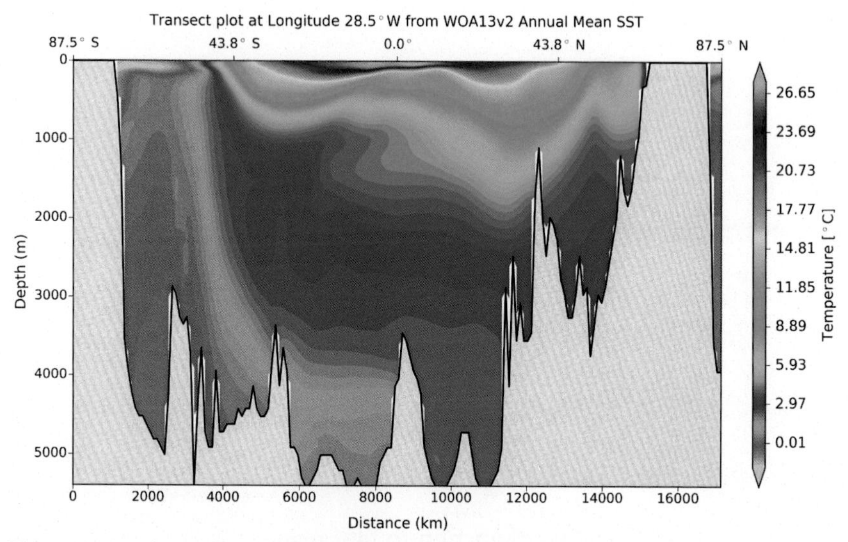

Fig. 4.11 A transect plot along longitude 28.5 °W in Python.

```
import numpy as np
import matplotlib.pyplot as plt
import matplotlib.ticker as ticker
from matplotlib.ticker import FuncFormatter
import cartopy.crs as ccrs
import cartopy.feature as cfeature
from cartopy.mpl.gridliner import LONGITUDE_FORMATTER,
LATITUDE_FORMATTER
from ctd import plot_section
from oceans.colormaps import cm
from iris.pandas import as_data_frame
```

Define custom woa_subset function to load NetCDF data file and slice data at given coordinates

```
def woa_subset(url, bbox=[-179.875, -89.875, 179.875, 89.875],
level=None):
    cubes = iris.load_raw(url)

    for idx, cube in enumerate(cubes):
        if 'Objectively analyzed mean fields for' in cube.
long_name:
            cube = cubes.pop(idx)
            break
```

Continued

```
    # Subset data
    lon = iris.Constraint(longitude=lambda lon: bbox[0] <=
lon <= bbox[2])
    lat = iris.Constraint(latitude=lambda lat: bbox[1] <=
lat <= bbox[3])
    if level:
        dep = iris.Constraint(depth=lambda z: z == level)
        cube = cube.extract(lon & lat & dep)
    else:
        cube = cube.extract(lon & lat)

    if cube.ndim >= 3 and cube.shape[0] == 1: # Squeeze time
dimension.
        cube = cube[0, ...]
    return cube
```

Define custom make_map function to plot map

```
def make_map(projection=ccrs.PlateCarree(), resolution='110m'):
    fig, ax = plt.subplots(figsize=(11, 7), subplot_kw=dict
(projection=projection))
    ax.set_global()
    ax.coastlines(resolution=resolution, color='k')
    gl = ax.gridlines(draw_labels=True)
    gl.xlabels_top = gl.ylabels_right = False
    gl.xformatter = LONGITUDE_FORMATTER
    gl.yformatter = LATITUDE_FORMATTER
    ax.add_feature(cfeature.LAND, facecolor='0.75')
    return fig, ax
```

Define data subsetting coordinates and NetCDF file path
Coordinates are defined in the order of min-lon, min-lat, max-lon, max-lat within bbox

```
ocean = dict(ww=dict(bbox=[-28.5, -87.5, -28.5, 87.5]))

level = None
ocean = ocean['ww']

base = '/path-to-local-data-folder/'
var = dict(temp='woa13_decav_t00_01v2.nc')
```

Start data subsetting and plotting

```
for key, value in var.items():
    print(key)
    url = base + value
    fname = value.split('/')[-1]
    cube = woa_subset(url, bbox=ocean['bbox'], level=level)
    if cube.ndim > 2:
        cube = cube.collapsed(['longitude'], iris.analysis.
MEAN)

    if key == 'temp':
        title = 'Temperature'
        units = r'$^\circ$C'
        cmap = cm.avhrr
 elif key == 'sal':
        title = 'Salinity'
        units = r'g kg$^{-1}$'
        cmap = cm.odv
 elif key == 'nit':
        title = 'Nitrogen'
        units = r'$\mu$mol L$^{-1}$'
        cmap = cm.odv
 elif key == 'aou' or key == 'oxy':
        title = 'AOU'
        units = r'mL L$^{-1}$'
        cmap = cm.odv
 elif key == 'sil':
        title = 'Silicate'
        units = r'$\mu$mol L$^{-1}$'
        cmap = cm.odv
 elif key == 'phos':
        title = 'Phosphate'
        units = r'$\mu$mol L$^{-1}$'
        cmap = cm.odv

 lon = cube.coord(axis='Y').points
 lat = cube.coord(axis='X').points

 df = as_data_frame(cube)
 if key == 'sal':
     df = df.clip(33, 37) # 27.5, 37.5
 levels = np.arange(df.min().min(), df.max().max(), 0.02)
 df.lat = lat
 df.lon = np.repeat(lon, len(lat))
```

Continued

```
    fig, ax, cb = plot_section(df, cmap=cm.odv, marker=None,
levels=levels, aspect=30, shrink=1.0, fraction=0.15)
    cb.set_label('Temperature [$^\circ$C]')
    cb.ax.tick_params(direction='out')
    ax.set_ylabel('Depth (m)')
    ax.set_xlabel('')
    def cb_ticklabel(x,pos):
        return "{:.2f}".format(x)

    cb.formatter = ticker.FuncFormatter(cb_ticklabel)
    cb.update_ticks()

    # New x-axis.
    def deg(labels):
        new_labels = []
        for x in labels:
            if x < 0:
                text = r'%.1f$^\circ$ S' % abs(x)
            elif x > 0:
                text = r'%.1f$^\circ$ N' % x
            else:
                text = r'%.1f$^\circ$' % x
            new_labels.append(text)
        return new_labels

    ax2 = ax.twiny()
    xmin, xmax = ax.get_xlim()[0], ax.get_xlim()[-1]
    loc = np.array([xmin, xmax*0.25, xmax*0.5, xmax*0.75, xmax])
    labels = [np.percentile(lon, 0), np.percentile(lon, 25), np.
percentile(lon, 50), np.percentile(lon, 75), np.percentile(lon,
100)]
    ax2.set_xticks(loc)
    ax2.set_xticklabels(deg(labels), fontdict={'fontsize':11})
    ax2.set_xlabel(r'Transect plot for WOA13v2 Annual Mean SST
at Longitude 28.5$^\circ$W', labelpad=10)
    ax2.tick_params(direction='out')
    ymax = ax.get_ylim()[0]
    plt.text(xmax/2-xmax*0.075,ymax*1.1,'Distance (km)')
```

Save and show figure

```
plt.savefig('/path-to-figure-folder/figure-name.png',
format='png', dpi=600, transparent=False)

plt.show()
```

4.12 Surface plots in Python

Surface plots can be generated in Python using plt.contourf and plt.pcolormesh functions from matplotlib library or using iplt.contourf function from iris library. In this section, six different examples are given for surface plots in Python.

Example 1

In this example, World Ocean Atlas (WOA) 2013v2 Annual Mean Temperature data [3] is used. WOA13v2 data is provided by the NOAA/NCEI OCL, USA, from their website at https://www.nodc.noaa.gov/OC5/woa13/ [4].

WOA13v2 1° grid annual temperature data can be downloaded from:

https://data.nodc.noaa.gov/thredds/fileServer/woa/WOA13/DATAv2/temperature/netcdf/decav/1.00/woa13_decav_t00_01v2.nc

This example (Fig. 4.12) is a modified version of the original source code provided at Cartopy documentation website (https://scitools.org.uk/cartopy/docs/v0.15/matplotlib/advanced_plotting.html) [7].

Key points in this surface plot:

- WOA13v2 1° grid annual temperature data is loaded with netCDF4 library
- basemap plotted with cartopy library
- data plotted with plt.contourf function from matplotlib library

A surface plot for worldwide annual mean SST data from WOA13v2 database:

Load required libraries

```
import os
import matplotlib.pyplot as plt
from netCDF4 import Dataset as netcdf_dataset
import numpy as np
from cartopy import config
import cartopy.crs as ccrs
from cartopy.util import add_cyclic_point
```

Define variables and load data

```
file='/path-to-local-data-folder/woa13_decav_t00_01v2.nc'

dataset = netcdf_dataset(file)
sst = dataset.variables['t_an'][0,0,:,:]
lats = dataset.variables['lat'][:]
lons = dataset.variables['lon'][:]
depth= dataset.variables['depth'][:]
time= dataset.variables['time'][:]
```

Continued

Example 1—cont'd

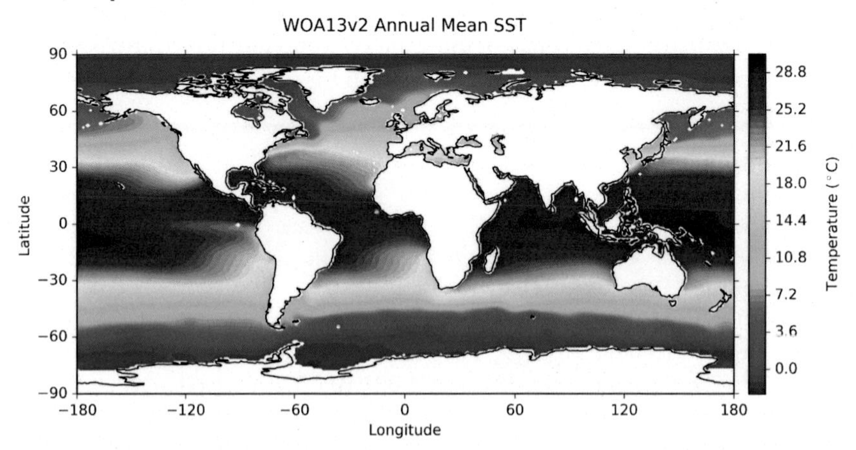

Fig. 4.12 A surface plot for worldwide annual mean SST data from WOA13v2 database in Python with NetCDF4 and Cartopy libraries.

Add cyclic_point to plot data on 0,0,:,:]

```
sst, lons = add_cyclic_point(sst, coord=lons)
```

Define map projection with ccrs.PlateCarree function in Cartopy
Plot data with plt.contourf function
Plot colorbar with plt.colorbar function

```
fig = plt.figure(figsize=(10, 6))

proj = ccrs.PlateCarree()

ax = plt.axes(projection=proj)

cf = plt.contourf(lons, lats, sst, 60, norm=None,
transform=proj)
cb = plt.colorbar(cf, extend='both', shrink=0.675,
pad=0.02, orientation='vertical', fraction=0.1)
cb.ax.set_ylabel('Temperature [$^\circ$C]')
cb.ax.get_yaxis().labelpad = 15
ax.coastlines()
ax.set_xticks([-180, -120, -60, 0, 60, 120, 180])
ax.set_yticks([-90, -60, -30, 0, 30, 60, 90])
ax.set_xlabel('Longitude')
ax.set_ylabel('Latitude')
ax.tick_params(direction='out')
cb.ax.tick_params(direction='out')
plt.title('WOA13v2 Annual Mean SST', y=1.05)
```

Set plot margins to tight layout; save and show figure

```
plt.tight_layout()

plt.savefig('/path-to-figure-folder/figure-name.png',
format='png',  dpi=600, transparent=False)

plt.show()
```

Example 2

In this example, World Ocean Atlas (WOA) 2009 Annual Mean Temperature data [8] is used. WOA09 data is provided by the NOAA/NCEI OCL, USA, from their website at https://www.nodc.noaa.gov/OC5/WOA09/pr_woa09.html [9].

WOA09 1° grid annual temperature data can be downloaded from:

http://data.nodc.noaa.gov/thredds/fileServer/woa/WOA09/NetCDFdata/temperature_annual_1deg.nc

This example (Fig. 4.13) is a modified version of the original source code provided at Cartopy documentation website (https://scitools.org.uk/cartopy/docs/v0.15/matplotlib/advanced_plotting.html) [7].

Key points in this surface plot:

- WOA09 1° grid annual temperature data is loaded with netCDF4 library
- basemap plotted with cartopy library
- data plotted with plt.contourf function from matplotlib library

A surface plot for worldwide annual mean SST data from WOA09 database

Load required libraries

```
import os
import matplotlib.pyplot as plt
from netCDF4 import Dataset as netcdf_dataset
import numpy as np
from cartopy import config
import cartopy.crs as ccrs
from cartopy.util import add_cyclic_point
```

Continued

Example 2—cont'd

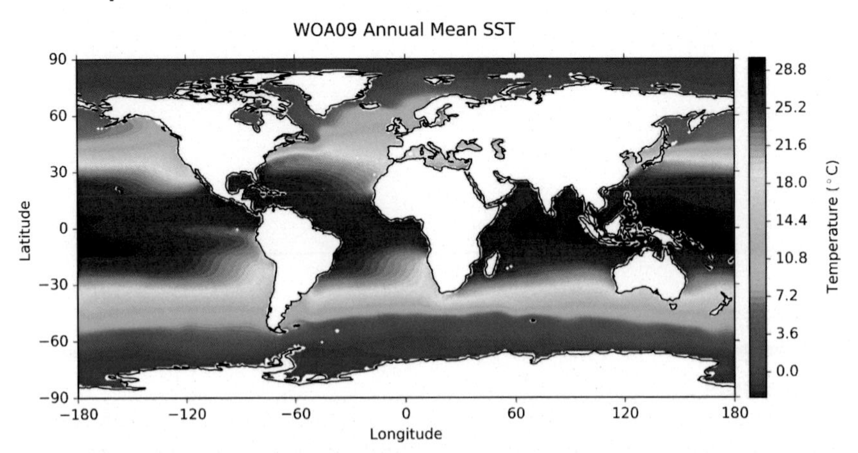

Fig. 4.13 A surface plot for worldwide annual mean SST data from WOA09 database in Python with NetCDF4 and Cartopy libraries.

Define variables and load data

```
file = '/path-to-local-data-
folder/temperature_annual_1deg.nc'

dataset = netcdf_dataset(file)
sst = dataset.variables['t_an'][0,0,:,:]
lats = dataset.variables['lat'][:]
lons = dataset.variables['lon'][:]
depth= dataset.variables['depth'][:]
time= dataset.variables['time'][:]
```

Add cyclic point to plot data on 0 longitude

```
sst, lons = add_cyclic_point(sst, coord=lons)
```

Define map projection with ccrs.PlateCarree function in Cartopy
Plot data with plt.contourf function
Plot colorbar with plt.colorbar function

```
fig = plt.figure(figsize=(10, 6))

proj = ccrs.PlateCarree()

ax = plt.axes(projection=proj)
```

```
cf = plt.contourf(lons, lats, sst, 60, norm=None,
transform=proj)
cb = plt.colorbar(cf, extend='both', shrink=0.675,
pad=0.02, orientation='vertical', fraction=0.1)
cb.ax.set_ylabel('Temperature [$^\circ$C]')
cb.ax.get_yaxis().labelpad = 15
ax.coastlines()
ax.set_xticks([-180, -120, -60, 0, 60, 120, 180])
ax.set_yticks([-90, -60, -30, 0, 30, 60, 90])
ax.set_xlabel('Longitude')
ax.set_ylabel('Latitude')
ax.tick_params(direction='out')
cb.ax.tick_params(direction='out')
plt.title('WOA09 Annual Mean SST', y=1.05)
```

Set plot margins to tight layout; save and show figure

```
plt.tight_layout()

plt.savefig('/path-to-figure-folder/figure-name.png',
format='png', dpi=600, transparent=False)

plt.show()
```

Example 3

In this example, NOAA OI SST V2 High Resolution Daily Mean 2018 data [10] is used. NOAA OI SST V2 High Resolution Daily Mean 2018 data is provided by the NOAA/OAR/ESRL PSD, Boulder, Colorado, USA, from their website at https://www.esrl.noaa.gov/psd/data/gridded/data.noaa.oisst.v2.highres.html [11].

NOAA OI SST V2 High Resolution Daily Mean 2018 data can be downloaded from:

ftp://ftp.cdc.noaa.gov/Datasets/noaa.oisst.v2.highres/sst.day.mean.2018.nc

This example (Fig. 4.14) is a modified version of the original source code provided at Cartopy documentation website (https://scitools.org.uk/cartopy/docs/v0.15/matplotlib/advanced_plotting.html) [7].

Key points in this surface plot:

- NOAA OI SST V2 High Resolution Daily Mean 2018 data is loaded with netCDF4 library
- basemap plotted with cartopy library
- data plotted with plt.contourf function from matplotlib library

Continued

Example 3—cont'd

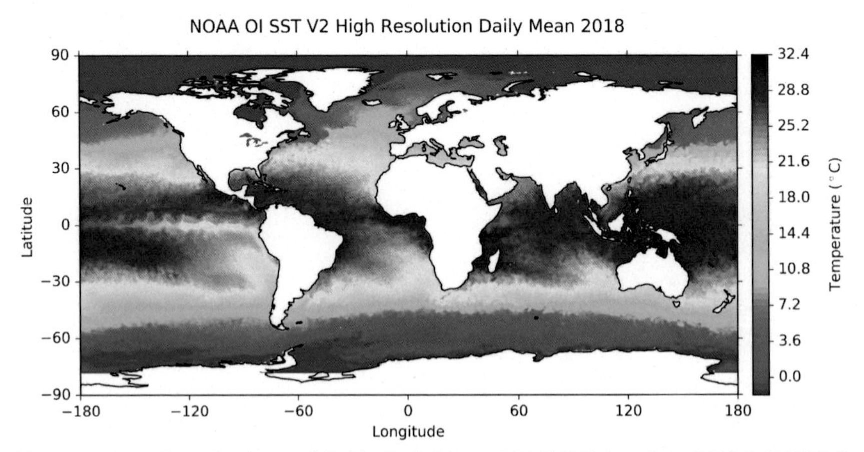

Fig. 4.14 A surface plot for worldwide Daily Mean (2018) SST data from NOAA OI SST V2 High Resolution database in Python with NetCDF4 and Cartopy libraries.

A surface plot for worldwide Daily Mean (2018) SST data from NOAA OI SST V2 High Resolution database:
Load required libraries

```
import os
import matplotlib.pyplot as plt
from netCDF4 import Dataset as netcdf_dataset
import numpy as np
from cartopy import config
import cartopy.crs as ccrs
from cartopy.util import add_cyclic_point
```

Define variables and load data

```
file='/path-to-local-data-folder/ sst.day.mean.2018.nc'

dataset = netcdf_dataset(file)
sst = dataset.variables['sst'][0, :, :]
lats = dataset.variables['lat'][:]
lons = dataset.variables['lon'][:]
```

Add cyclic point to plot data on 0 longitude

```
sst, lons = add_cyclic_point(sst, coord=lons)
```

Define map projection with ccrs.PlateCarree function in Cartopy
Plot data with plt.contourf function
Plot colorbar with plt.colorbar function

```
fig = plt.figure(figsize=(10, 6))

proj = ccrs.PlateCarree()

ax = plt.axes(projection=proj)

cf = plt.contourf(lons, lats, sst, 60, norm=None,
transform=proj)

cb = plt.colorbar(cf, extend='both', shrink=0.675,
pad=0.02, orientation='vertical', fraction=0.1)

cb.ax.set_ylabel('Temperature [$o\circ$C]')
cb.ax.get_yaxis().labelpad = 15
ax.coastlines()
ax.set_xticks([-180, -120, -60, 0, 60, 120, 180])
ax.set_yticks([-90, -60, -30, 0, 30, 60, 90])
ax.set_xlabel('Longitude')
ax.set_ylabel('Latitude')
ax.tick_params(direction='out')
cb.ax.tick_params(direction='out')
plt.title('NOAA OI SST V2 High Resolution Daily Mean
2018', y=1.05)
```

Set plot margins to tight layout; save and show figure

```
plt.tight_layout()

plt.savefig('/path-to-figure-folder/figure-name.png',
format='png', dpi=600, transparent=False)

plt.show()
```

Example 4

In this example, World Ocean Atlas (WOA) 2013v2 Annual Mean Temperature data [3] is used. WOA13v2 data is provided by the NOAA/NCEI OCL, USA, from their website at https://www.nodc.noaa.gov/OC5/woa13/ [4].

WOA13v2 1° grid annual temperature data can be downloaded from:

https://data.nodc.noaa.gov/thredds/fileServer/woa/WOA13/DATAv2/temperature/netcdf/decav/1.00/woa13_decav_t00_01v2.nc

This example (Fig. 4.15) is a modified version of the original source code provided by Filipe Pires Alvarenga Fernandes at python4oceanographers website (https://ocefpaf.github.io/python4oceanographers/blog/2013/09/16/iris/) [12]. The original source code is licensed under a Creative Commons Attribution-ShareAlike 4.0 International License.

Key points in this surface plot:

- WOA13v2 1° grid annual temperature data is loaded with cubes from iris library
- basemap plotted with cartopy library
- data plotted with iplt.contourf function from iris library

A surface plot for worldwide annual mean SST data from WOA13v2 database:

Load required libraries

```
import numpy as np
import matplotlib.pyplot as plt
from oceans.colormaps import cm
import cartopy.crs as ccrs
import iris
import iris.plot as iplt
from iris.fileformats.netcdf import load_cubes
iris.FUTURE.netcdf_promote=True
```

Define data path, variables, and load data

```
url = '/path-to-local-data-folder/woa13_decav_t00_01v2.
nc'
cubes = dict()
for cube in load_cubes([url]):
    cubes.update({cube.long_name: cube})

print(cubes.keys())

name = u'Objectively analyzed mean fields for
sea_water_temperature at standard depth levels.'

sst=cubes[name].slices(['latitude', 'longitude']).next()
```

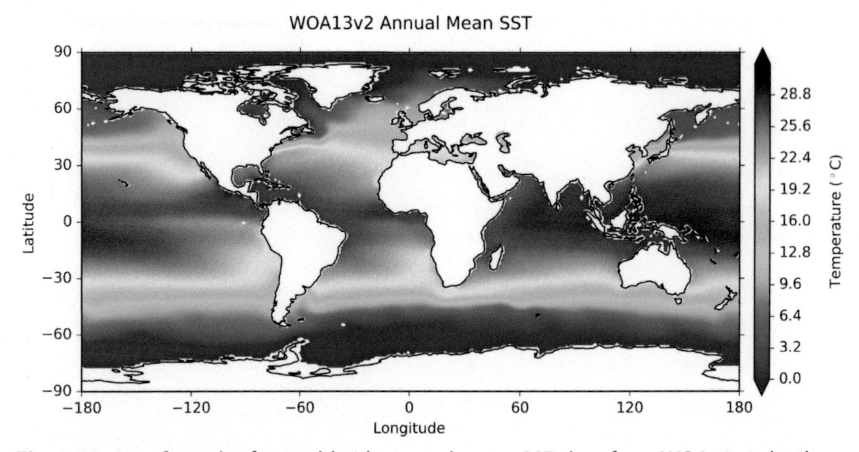

Fig. 4.15 A surface plot for worldwide annual mean SST data from WOA13v2 database in Python with iris and Cartopy libraries.

Define contour levels
Define map projection with ccrs.PlateCarree function in Cartopy
Plot data with iplt.contourf function
Plot colorbar with fig.colorbar function

```
levels = np.arange(0, 32, 0.05)

fig = plt.figure(figsize=(10, 6))

proj = ccrs.PlateCarree()

ax = plt.axes(projection=proj)

cs = iplt.contourf(sst, cmap=cm.avhrr, levels=levels,
extend='both')

ax.coastlines()
ax.set_xticks([-180, -120, -60, 0, 60, 120, 180])
ax.set_yticks([-90, -60, -30, 0, 30, 60, 90])
ax.set_xlabel('Longitude')
ax.set_ylabel('Latitude')

cb   =   fig.colorbar(cs,   extend='both',   shrink=0.69,
pad=0.02, orientation='vertical', fraction=0.1)

cb.ax.get_yaxis().labelpad = 15

cb.set_label('Temperature [$o\circ$C]', rotation=270)

ax.tick_params(direction='out')
```

Continued

Example 4—cont'd

```
cb.ax.tick_params(direction='out')
plt.title('WOA13v2 Annual Mean SST', y=1.05)
```

Set plot margins to tight layout; save and show figure

```
plt.tight_layout()
plt.savefig('/path-to-figure-folder/figure-name.png',
format='png', dpi=600, transparent=False)
plt.show()
```

Example 5

In this example, World Ocean Atlas (WOA) 2009 Annual Mean Temperature data [8] is used. WOA09 data is provided by the NOAA/NCEI OCL, USA, from their website at https://www.nodc.noaa.gov/OC5/WOA09/pr_woa09.html [9].

WOA09 1° grid annual temperature data can be downloaded from: http://data.nodc.noaa.gov/thredds/fileServer/woa/WOA09/NetCDFdata/temperature_annual_1deg.nc

This example (Fig. 4.16) is a modified version of the original source code provided by Filipe Pires Alvarenga Fernandes at python4oceanographers website (https://ocefpaf.github.io/python4oceanographers/blog/2013/09/16/iris/) [12]. The original source code is licensed under a Creative Commons Attribution-ShareAlike 4.0 International License.

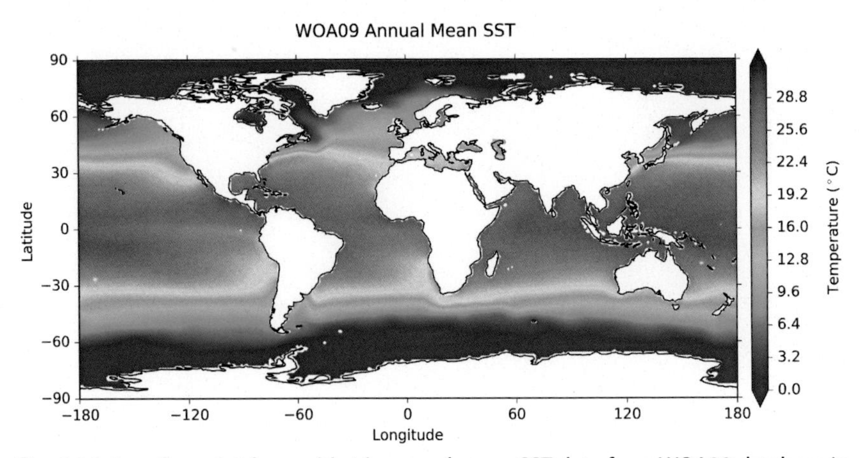

Fig. 4.16 A surface plot for worldwide annual mean SST data from WOA09 database in Python with iris and Cartopy libraries.

Continued

Key points in this surface plot:

- WOA09 1° grid annual temperature data is loaded with cubes from iris library
- basemap plotted with cartopy library
- data plotted with iplt.contourf function from iris library

A surface plot for worldwide annual mean SST data from WOA09 database:

Load required libraries

```
import numpy as np
import matplotlib.pyplot as plt
from oceans.colormaps import cm
import cartopy.crs as ccrs
import iris
import iris.plot as iplt
from iris.fileformats.netcdf import load_cubes
iris.FUTURE.netcdf_promote=True
```

Define data path, variables and load data

```
url = '/path-to-local-data-
folder/temperature_annual_1deg.nc'
cubes = dict()
for cube in load_cubes([url]):
    cubes.update({cube.long_name: cube})

print(cubes.keys())

name = u'Objectively Analyzed Climatology'

sst=cubes[name].slices(['latitude', 'longitude']).next()
```

Define contour levels
Define map projection with ccrs.PlateCarree function in Cartopy
Plot data with iplt.contourf function
Plot colorbar with fig.colorbar function

```
levels = np.arange(0, 32, 0.05)

fig = plt.figure(figsize=(10, 6))

proj = ccrs.PlateCarree()
```

Continued

Example 5—cont'd

```
ax = plt.axes(projection=proj)

cs = iplt.contourf(sst, cmap=cm.avhrr, levels=levels,
extend='both')

ax.coastlines()
ax.set_xticks([-180, -120, -60, 0, 60, 120, 180])
ax.set_yticks([-90, -60, -30, 0, 30, 60, 90])
ax.set_xlabel('Longitude')
ax.set_ylabel('Latitude')

cb = fig.colorbar(cs, extend='both', shrink=0.71,
pad=0.02, orientation='vertical', fraction=0.1)

cb.ax.get_yaxis().labelpad = 15

cb.set_label('Temperature ($^\circ$C)')

ax.tick_params(direction='out')

cb.ax.tick_params(direction='out')

plt.title('WOA09 Annual Mean SST', y=1.05)
```

Set plot margins to tight layout; save and show figure

```
plt.tight_layout()

plt.savefig('/path-to-figure-folder/figure-name.png',
format='png', dpi=600, transparent=False)

plt.show()
```

Example 6

In this example, World Ocean Atlas (WOA) 2009 Annual Mean Temperature data [8] is used. WOA09 data is provided by the NOAA/NCEI OCL, USA, from their website at https://www.nodc.noaa.gov/OC5/WOA09/pr_woa09.html [9].

WOA09 1° grid annual temperature data can be downloaded from:

http://data.nodc.noaa.gov/thredds/fileServer/woa/WOA09/NetCDFdata/temperature_annual_1deg.nc

"Blue Marble: Next Generation" image is provided by the NASA Earth Observations (NEO), USA, from their website at https://neo.sci.gsfc.nasa.gov/ [13]. Background image file can be downloaded from: https://neo.sci.gsfc.nasa.gov/view.php?datasetId=BlueMarbleNG-TB

This example (Fig. 4.17) is a modified version of the original source code provided by Nikolay Koldunov at EarthPy website (http://earthpy.org/cartopy_backgroung.html) [14].

Key points in this surface plot:

- WOA09 1° grid annual temperature data is loaded with netCDF4 library
- basemap plotted with cartopy library
- Blue Marble image is used as map background with ax.background_img function from Cartopy library
- data plotted with plt.pcolormesh function from matplotlib library

In order to use an image as a map background, a json file with below codes should be created and optionally this file can be placed into the cartopy folder of your python environment. In the example, the json file is placed under cartopy/bg path.

Save below codes as a json file with "images" name:

```
{"__comment__": "JSON file specifying the image to use
for a given name and resolution",
  "bmarb": {
    "__comment__":  "Blue  Marble:  Next  Generation
+Topography and Bathymetry, December 2004 ",
    "__source__":
"https://neo.sci.gsfc.nasa.gov/view.php?
datasetId=BlueMarbleNG-TB&date=2004-12-01",
    "__projection__": "PlateCarree",
    "medium": "bmarb_med.png",
    "high": "bmarb_high.png"}
}
```

In this json file, 0.25 degrees and 1440x720 resolution png file are used as medium resolution image, and 0.1 degrees and 3600x1800 resolution png file are used as high resolution image.

Continued

Example 6—cont'd

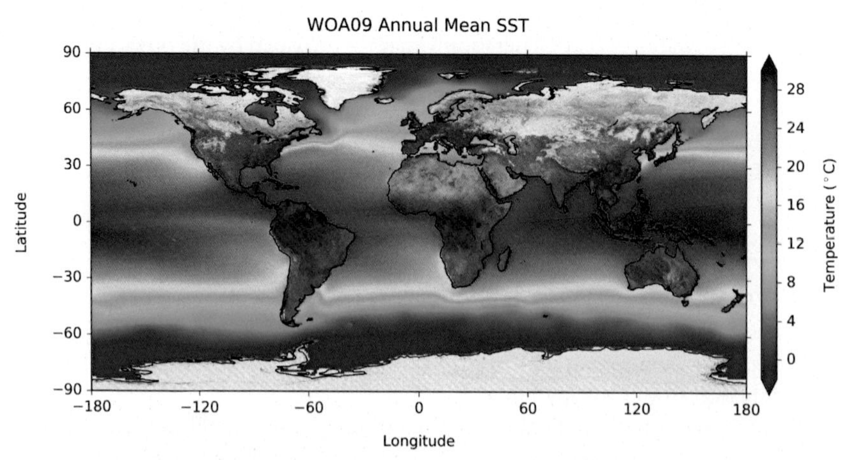

Fig. 4.17 A surface plot for worldwide annual mean SST data from WOA09 database in Python with Blue Marble background.

A surface plot for worldwide annual mean SST data from WOA09 database with Blue Marble background:

Load required libraries

```
import os
import sys
import numpy as np
import matplotlib.pylab as plt
import cartopy.crs as ccrs
from cartopy.util import add_cyclic_point
from netCDF4 import MFDataset, num2date
from netCDF4 import Dataset as netcdf_dataset
from oceans.colormaps import cm
```

Define json file data path, variables and load data

```
os.environ['CARTOPY_USER_BACKGROUNDS'] = '/path-to-
python-environemnt/site-packages/cartopy/bg/'

file = '/path-to-local-data-
folder/temperature_annual_1deg.nc'

dataset = netcdf_dataset(file)
lat = dataset.variables['lat'][:]
lon = dataset.variables['lon'][:]
sst = dataset.variables['t_an'][0,0,:,:]
```

Add cyclic point to plot data on 0 longitude

```
sst, lon = add_cyclic_point(sst, coord=lon)
```

Define map projection with ccrs.PlateCarree function in Cartopy
Add high resolution background image with ax.background_img function from Cartpoy
Plot data with plt.pcolormesh function
Plot colorbar with plt.colorbar function

```
fig = plt.figure(figsize=(10,6))

proj = ccrs.PlateCarree()

ax = plt.axes(projection=proj)

ax.background_img(name='bmarb', resolution='high')

cs = plt.pcolormesh(lon, lat, sst, vmin=-2, vmax=30,
transform=proj, cmap=cm.avhrr)

ax.coastlines()

cb = plt.colorbar(cs, extend='both', shrink=0.675,
pad=0.02, orientation='vertical', fraction=0.1)

cb.ax.get_yaxis().labelpad = 15
cb.set_label('Temperature ($^\circ$C)')
ax.set_xticks([-180, -120, -60, 0, 60, 120, 180])
ax.set_yticks([-90, -60, -30, 0, 30, 60, 90])
ax.set_xlabel('Longitude', labelpad=15)
ax.set_ylabel('Latitude', labelpad=20)
ax.tick_params(direction='out')
cb.ax.tick_params(direction='out')
plt.title('WOA09 Annual Mean SST', y=1.05)
```

Set plot margins to tight layout; save and show figure

```
plt.tight_layout()

plt.savefig('/path-to-figure-folder/figure-name.png',
format='png', dpi=600, transparent=False)

plt.show()
```

4.13 Animations in R and Python

Animations are frequently used in representation of model outputs, and they can be generated by using packages/libraries with specialized functions for animations or by saving plots as multiple images and converting these images into gif or video files.

In R, animations can be prepared with packages like animation and gganimate. In Python, animations can be created with animation module of matplotlib, MoviePy library, or with other libraries like Pyglet.

Easiest way for producing animations is to generate plots by using loops and saving them as image files such as jpeg or png. Then, these images can be converted to gifs or movie formats (such as mp4) by using tools like ImageMagick and ffmpeg.

In this section, you will find a simple animation written with animation module of matplotlib. This simple example (Fig. 4.18) can be used to visualize sampling routes of a research vessel and sampling stations on a map.

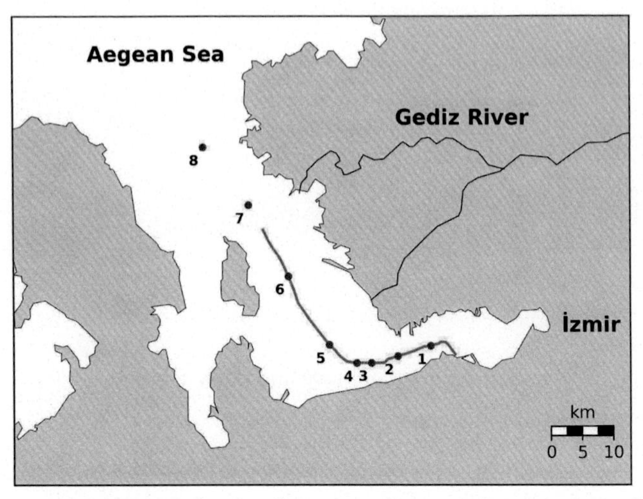

Fig. 4.18 A simple animation for visualizing sampling routes and stations of a research vessel on a map in Python.

Coordinate data used in this example is given below. Save this data as txt file to run the example:

```
lat        lon
38.4153    27.0361
38.4198    27.0319
38.4218    27.0299
38.4260    27.0259
38.4286    27.0222
38.4291    27.0182
38.4290    27.0162
38.4286    27.0142
38.4273    27.0125
38.4261    27.0061
38.4248    26.9999
38.4240    26.9915
38.4215    26.9830
38.4194    26.9774
38.4162    26.9652
38.4133    26.9524
38.4102    26.9386
38.4060    26.9321
38.4058    26.9248
38.4057    26.9171
38.4057    26.9137
38.4059    26.9103
38.4058    26.9066
38.4058    26.9029
38.4058    26.8991
38.4057    26.8919
38.4072    26.8777
38.4121    26.8691
38.4146    26.8655
38.4193    26.8607
38.4259    26.8518
38.4388    26.8380
38.4556    26.8214
38.4724    26.8065
38.4867    26.8004
38.5021    26.7922
38.5209    26.7811
38.5390    26.7647
38.5534    26.7556
38.5670    26.7450
38.5806    26.7340
```

Continued

—cont'd

38.5896	26.7213
38.6021	26.7055
38.6155	26.6916
38.6315	26.6786
38.6449	26.6672
38.6567	26.6520
38.6724	26.6348
38.6867	26.6248
38.7023	26.6211

Load required libraries

```
import numpy as np
import matplotlib.pyplot as plt
import matplotlib.animation as animation
from mpl_toolkits.basemap import Basemap
import os
os.chdir('/path-of-working-directory/')
```

Load coordinate data

```
# plot points of acceleration and deceleration
f = open('data.txt')
cr = np.loadtxt(f, unpack=True, usecols=None, skiprows=1,
delimiter='\t')

x = cr[1]
y = cr[0]
```

Select sampling stations from coordinate data by using i th coordinate in "ind" variable
"st" variable is created to serve as an index variable in the loop

```
ind = [10, 15, 20, 25, 30, 35, 40, 45]
st = [1, 2, 3, 4, 5, 6, 7, 8]

sx = x[[ind]]
sy = y[[ind]]
```

Define minimum and maximum map limits

```
xmin = 26.39
ymin = 38.27
xmax = 27.29
ymax = 38.79
```

```
# Plot the map
```

```
fig, ax = plt.subplots()

m = Basemap(projection = 'merc', llcrnrlon = xmin, llcrnrlat =
ymin, urcrnrlon = xmax, urcrnrlat = ymax, resolution='h')

m.drawcoastlines(color='black',linewidth=0.5)

m.drawmapboundary(fill_color='white')

m.fillcontinents(color='lightgrey',lake_color='white')

m.drawrivers(linewidth=0.8, linestyle='solid', color='k',
antialiased=1, ax=None, zorder=None)
```

```
# Convert variables into map coordinates by using m as a function
```

```
xmin, ymin = m(xmin, ymin)
xmax, ymax = m(xmax, ymax)
stx, sty = m(sx-0.020,sy-0.020)
sx,sy = m(sx, sy)
x,y = m(x, y)
```

```
# Define parameters for line, points, and texts plots
```

```
line, = ax.plot(x, y, color='royalblue', ls='-', lw=2)
point = ax.plot(sx, sy, 'ro', markersize=5)[0]

kw = {'fontsize': 10, 'fontweight': 'bold'}
for ind, txt in enumerate(ind):
    ax.annotate(st[ind], (stx[ind], sty[ind]), **kw)

lon = [26.5, 26.95, 27.19]
lat = [38.74, 38.67, 38.44]
txtp = [u'Aegean Sea', u'Gediz River', u'⊠zmir']
pt = lon
v, z = m(lon, lat)
kw = {'fontsize': 15, 'fontweight': 'bold'}

for pt, txt in enumerate(pt):
    ax.annotate(txtp[pt], (v[pt],z[pt]), **kw)
```

```
# Add scale bar
```

```
m.drawmapscale(27.22, 38.33, 27.30, 38.33,
    10,
    units='km', fontsize=12,
    yoffset=None,
```

Continued

```
        barstyle='fancy', labelstyle='simple',
        fillcolor1='w', fillcolor2='#000000',
        fontcolor='#000000',
        zorder=5)
```

Define animated objects and set variables

```
def update(num, x, y, line):
    line.set_data(x[:num], y[:num])
    line.axes.axis([xmin, xmax, ymin, ymax])
    point.set_data(sx, sy)
    return line, point
```

Start animation

```
ani = animation.FuncAnimation(fig, update, len(x), fargs=[x, y,
line], interval=125, blit=True)
```

Save (or show) animation

```
ani.save('animation.mp4', fps=8, bitrate=128, dpi=300)

#plt.show()
```

References

[1] R for Science Website, Salinity page, Author: Karline Soetaert, http://www.rforscience.com/portfolio/salinity/. Accessed 17 August 2018.

[2] GSHHG Website, https://www.soest.hawaii.edu/pwessel/gshhg/. Accessed 17 August 2018.

[3] R.A. Locarnini, A.V. Mishonov, J.I. Antonov, T.P. Boyer, H.E. Garcia, O. K. Baranova, M.M. Zweng, C.R. Paver, J.R. Reagan, D.R. Johnson, M. Hamilton, D. Seidov, in: S. Levitus (Ed.), World Ocean Atlas 2013, Volume 1: Temperature, NOAA Atlas NESDIS 73, 2013. A. Mishonov Technical Ed. 40 pp.

[4] World Ocean Atlas (WOA) 2013v2 Website, https://www.nodc.noaa.gov/OC5/woa13/. Accessed 17 August 2018.

[5] R for Science Website, Silicate, phosphate and nitrate page, Author: Karline Soetaert, http://www.rforscience.com/portfolio/silicate-phosphate-and-nitrate/. Accessed 17 August 2018.

[6] Figures for lectures in oceanography website, Author: Juliana Leonel, https://juoceano.github.io/lecture_figures/. Accessed 17 August 2018.

[7] Cartopy documentation website, https://scitools.org.uk/cartopy/docs/v0.15/matplotlib/advanced_plotting.html. Accessed 17 August 2018.

[8] R.A. Locarnini, A.V. Mishonov, J.I. Antonov, T.P. Boyer, H.E. Garcia, O. K. Baranova, M.M. Zweng, D.R. Johnson, in: S. Levitus (Ed.), World Ocean Atlas 2009, Volume 1: Temperature, NOAA Atlas NESDIS 68, U.S. Government Printing Office, Washington, DC, 2010. 184 pp.

[9] World Ocean Atlas (WOA) Website, https://www.nodc.noaa.gov/OC5/WOA09/pr_woa09.html, 2009. Accessed 17 August 2018.

[10] R.W. Reynolds, T.M. Smith, C. Liu, D.B. Chelton, K.S. Casey, M.G. Schlax, Daily high-resolution-blended analyses for sea surface temperature, J. Climate 20 (2007) 5473–5496.

[11] NOAA OI SST V2 High Resolution Website, https://www.esrl.noaa.gov/psd/data/gridded/data.noaa.oisst.v2.highres.html. Accessed 17 August 2018.

[12] Python4oceanographers website, Author: Filipe Pires Alvarenga Fernandes, https://ocefpaf.github.io/python4oceanographers/blog/2013/09/16/iris/. Accessed 17 August 2018.

[13] "Blue Marble: Next Generation" image website, https://neo.sci.gsfc.nasa.gov/. Accessed 17 August 2018.

[14] EarthPy website, Author: Nikolay Koldunov, http://earthpy.org/cartopy_backgroung.html. Accessed 17 August 2018.

CHAPTER 5

Chemical oceanography examples

5.1 Vertical profiling plots in R

The following dataset will be used to produce vertical profiling plots in this chapter. There are 23 observations in the dataset and vertical Dissolved Oxygen (do), ortho-Phosphate (phos), Total Nitrite + Nitrate (tnox) and Silicate (si) data is given. Values do not belong to field observations.

	depth	do	phos	tnox	si
1	10	180	0.10	1.0	2.0
2	50	150	0.25	50.0	5.0
3	100	130	0.75	54.0	15.0
4	250	110	1.75	55.0	35.0
5	500	80	2.50	54.0	45.0
6	750	45	3.25	53.0	50.0
7	1000	50	3.50	52.0	55.0
8	1250	60	3.30	51.0	60.0
9	1500	70	3.20	50.0	60.5
10	1750	80	3.10	49.0	60.0
11	2000	90	3.00	48.0	59.5
12	2250	100	2.90	47.9	58.0
13	2500	110	2.80	47.6	56.0
14	2750	120	2.70	47.3	54.0
15	3000	130	2.60	47.0	52.0
16	3250	140	2.50	46.7	50.0
17	3500	146	2.40	46.4	48.0
18	3750	149	2.30	46.1	46.0
19	4000	152	2.20	45.8	44.0
20	4250	155	2.10	45.5	42.0
21	4500	158	2.05	45.2	41.0
22	4750	159	2.01	44.9	40.2
23	4900	160	1.99	44.6	39.8

A vertical profiling plot for a single variable (Fig. 5.1):

```
# Define working folder path and load data file
# Example path for Windows: "C:/Users/username/Documents/"
# Example path for Linux: "/home/username/Documents/"
# Example path for Mac OS: "/Users/username/Documents/"
```

R and Python for Oceanographers
https://doi.org/10.1016/B978-0-12-813491-7.00005-1

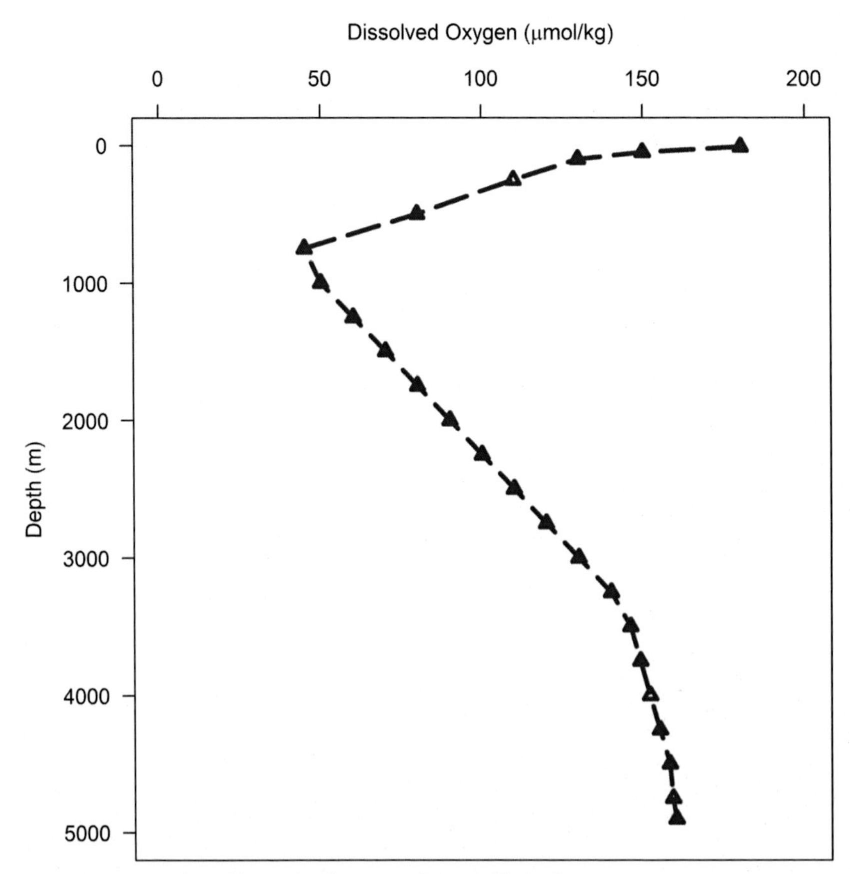

Fig. 5.1 A vertical profiling plot for a single variable in R.

```
setwd("/path-to-local-data-folder/")
data <- read.table("data.txt", header=TRUE)
attach(data)
```

Set axis ticks at defined intervals

```
x_ticks <- seq(0,200,50)
y_ticks <- seq(0,5000,1000)
```

Define output folder path for figure and set figure parameters

```
tiff("figure-name.tiff", width=14, height=14, unit="cm",
res=600,  pointsize=10, compression="lzw", bg="white")
par(mar=c(1,5,4,1), las=1, xpd=TRUE)
```

Define the plot
Note that the axes and lines are not drawn with plot function; they are plotted with axis and lines functions manually [1]
Axes are numbered as 1: bottom, 2: left, 3: top and 4: right.
To use top axis as x-axis, use 3 in axis function
To use left axis as y-axis, use 2 in axis function
Axis labels are drawn with mtext function; y axis label is rotated vertically using "las=0" in mtext function
Plot device is closed and saved with dev.off() function

```
plot(0, 0, type="o", xaxt="n", yaxt="n", xlim=c(0,200),
     ylim=rev(range(0,5000)), col="white", pch=0, cex=1,
     lwd=2,   main="",   xlab="", ylab="")
axis(2, at=y_ticks, labels=y_ticks, cex.axis=1)
axis(3, at=x_ticks, labels=x_ticks, cex.axis=1)
mtext(expression("Dissolved  Oxygen  ("*mu*"mol/kg)"),  side=3,
line=2.6, cex=1, las=1, col="black")
mtext("Depth (m)", side=2, line=3.5, cex=1, las=0,
col="black")
lines(data$do, data$depth, col="blue", pch=2, type="o",
lty="longdash", lwd=3)
dev.off()
```

A vertical profiling plot for two different variables plotted on the same axis (Fig. 5.2):

Define working folder path and load data file

```
setwd("/path-to-local-data-folder/")
data <- read.table("data.txt", header=TRUE)
attach(data)
```

Set axis ticks at defined intervals

```
x3_ticks <- seq(0,200,50)
x1_ticks <- seq(0,4,0.5)
y_ticks <- seq(5000,0,-1000)
```

Define output folder path for figure and set figure parameters

```
tiff("figure-name.tiff", width=14, height=14, unit="cm",
res=600,  pointsize=10, compression="lzw", bg="white")
par(mar=c(1,6,6,1), las=1, xpd=TRUE)
```

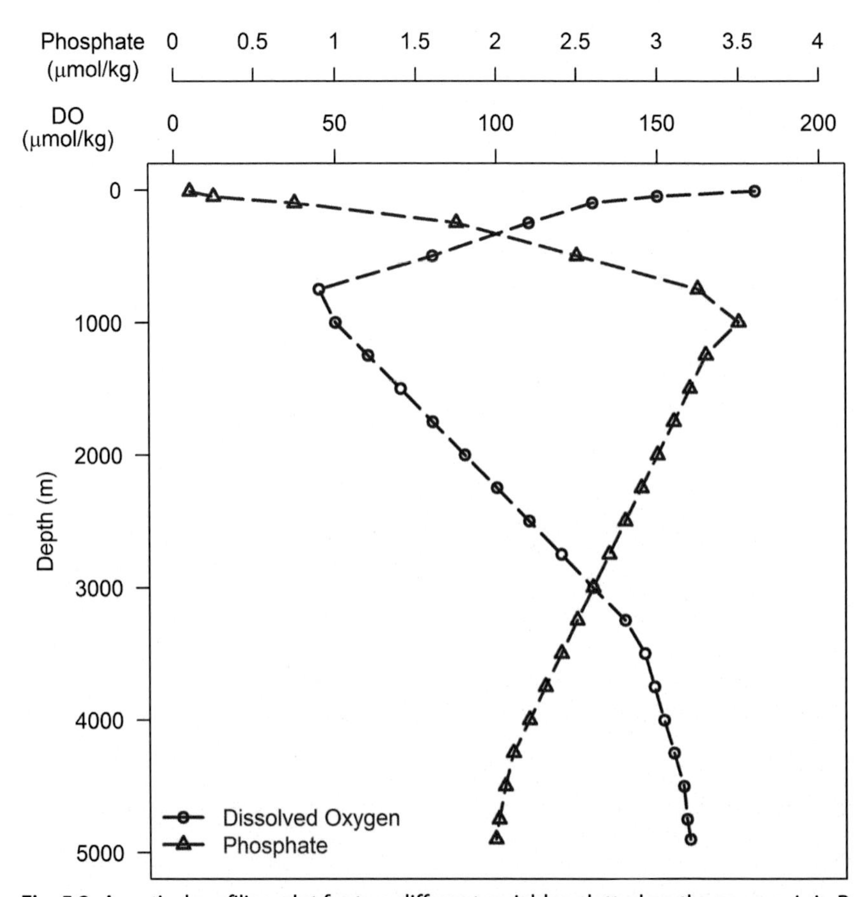

Fig. 5.2 A vertical profiling plot for two different variables plotted on the same axis in R.

Define the plot
Note that each axis is drawn by using plot and axis functions
Before the second plot function, par(new=T) is used to draw new data series after the first data series
To use top axis for two variables at different intervals and/or units, axis functions are used after each plot function
Figure legend is drawn with legend function

```
plot(data$do, data$depth, type="o", axes=FALSE,
     xlim=c(0,200),  ylim=rev(range(0,5000)), col="darkblue",
     pch=1, cex=1,  lty="longdash", lwd=2, main="", xlab="",
     ylab="")
axis(3, xlim=c(0,200), at=x3_ticks, labels=x3_ticks,
cex.axis=1, col="black", lwd=1)
```

```
mtext(3,text=expression("DO"), line=1.25, at=-33, cex=1)
mtext(3,text=expression("("*mu*"mol/kg)"), line=0.25, at=-33,
cex=1)
par(new=T)
plot(data$phos, data$depth, type="o", axes=FALSE,
     xlim=c(0,4), ylim=rev(range(0,5000)), col="red", pch=2,
     cex=1, lty="longdash", lwd=2, main="", xlab="",
     ylab="")
axis(3, xlim=c(0,4), at=x1_ticks, labels=x1_ticks, line=3,
cex.axis=1, col="black", lwd=1)
mtext(3, text="Phosphate", line=4, at=-0.5, cex=1)
mtext(3, text=expression("("*mu*"mol/kg)"), line=2.8, at=-
0.5, cex=1)
axis(2, at=y_ticks, labels=y_ticks, cex.axis=1)
mtext(2, text=" Depth (m)", line=3.5, cex=1, las=0)
box()
legend("bottomleft", legend=c("Dissolved
     Oxygen","Phosphate"), lty=1, lwd=2, pch=c(1,2),
     col=c("darkblue","red"), ncol=1, bty="n", cex=1,
     text.col=c("darkblue","red"), inset=0.01)
dev.off()
```

A vertical profiling plot for two different variables given on opposite axes (Fig. 5.3):

Define working folder path and load data file

```
setwd("/path-to-local-data-folder/")
data <- read.table("data.txt", header=TRUE)
attach(data)
```

Set axis ticks at defined intervals

```
x3_ticks <- seq(0,200,50)
x1_ticks <- seq(0,4,0.5)
y_ticks <- seq(5000,0,-1000)
```

Define output folder path for figure and set figure parameters

```
tiff("figure-name.tiff",   width=14,   height=14,   unit="cm",
res=600,  pointsize=10, compression="lzw", bg="white")
par(mar=c(4,5,4,1.5), las=1, xpd=TRUE)
```

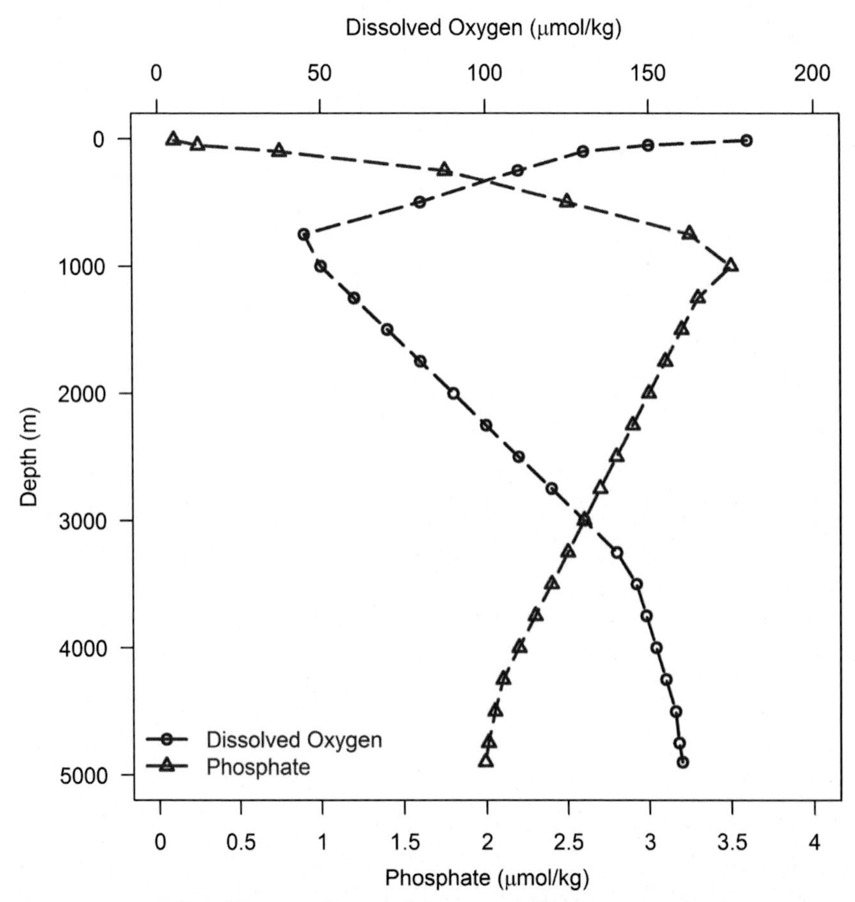

Fig. 5.3 A vertical profiling plot for two different variables given on opposite axes in R.

Define the plot
Note that each axis is drawn by using plot and axis functions
Before the second plot function, par(new=T) is used to draw new data series after the first data series

```
plot(data$do, data$depth, type="o", axes=FALSE,
     xlim=c(0,200),  ylim=rev(range(0,5000)), col="darkblue",
     pch=1, cex=1,  lty="longdash", lwd=2, main="", xlab="",
     ylab="")
axis(3, xlim=c(0,200), at=x3_ticks, labels=x3_ticks,
cex.axis=1, col="black", lwd=1)
mtext(3,text=expression("Dissolved Oxygen ("*mu*"mol/kg)"),
line=2.5, cex=1)
```

```
par(new=T)
plot(data$phos, data$depth, type="o", axes=FALSE,
     xlim=c(0,4), ylim=rev(range(0,5000)), col="red", pch=2,
     cex=1, lty="longdash", lwd=2, main="", xlab="",
     ylab="")
axis(1, xlim=c(0,4), at=x1_ticks, labels=x1_ticks,
cex.axis=1, col="black", lwd=1)
mtext(1, text=expression("Phosphate ("*mu*"mol/kg)"),
line=2.5, cex=1)
axis(2, at=y_ticks, labels=y_ticks, cex.axis=1)
mtext(2, text=" Depth (m)", line=3.5, cex=1, las=0)
box()
legend("bottomleft", legend=c("Dissolved
     Oxygen","Phosphate"), lty=1, lwd=2, pch=c(1,2),
     col=c("darkblue","red"), ncol=1, bty="n",  cex=1,
     text.col=c("darkblue","red"), inset=0.01)
dev.off()
```

A multipanel plot including 4 different subplots (Fig. 5.4):

Define working folder path and load data file
Define working folder

```
setwd("/path-to-local-data-folder/")
data <- read.table("data.txt", header=TRUE)
attach(data)
```

Set axis ticks at defined intervals

```
x1_ticks <- seq(0,200,50)
x2_ticks <- seq(0,4,0.5)
x3_ticks <- seq(0,60,10)
x4_ticks <- seq(0,70,10)
y_ticks <- seq(0,5000,1000)
```

Define output folder path for figure and set figure parameters
Order and number of subplots is set with mfcol or mfrow functions

```
tiff("figure-name.tiff",   width=28,   height=14,   unit="cm",
res=600,  pointsize=10, compression="lzw", bg="white")
par(mfcol=c(1,4), mar=c(1,6,4,1), las=1, xpd=TRUE)
```

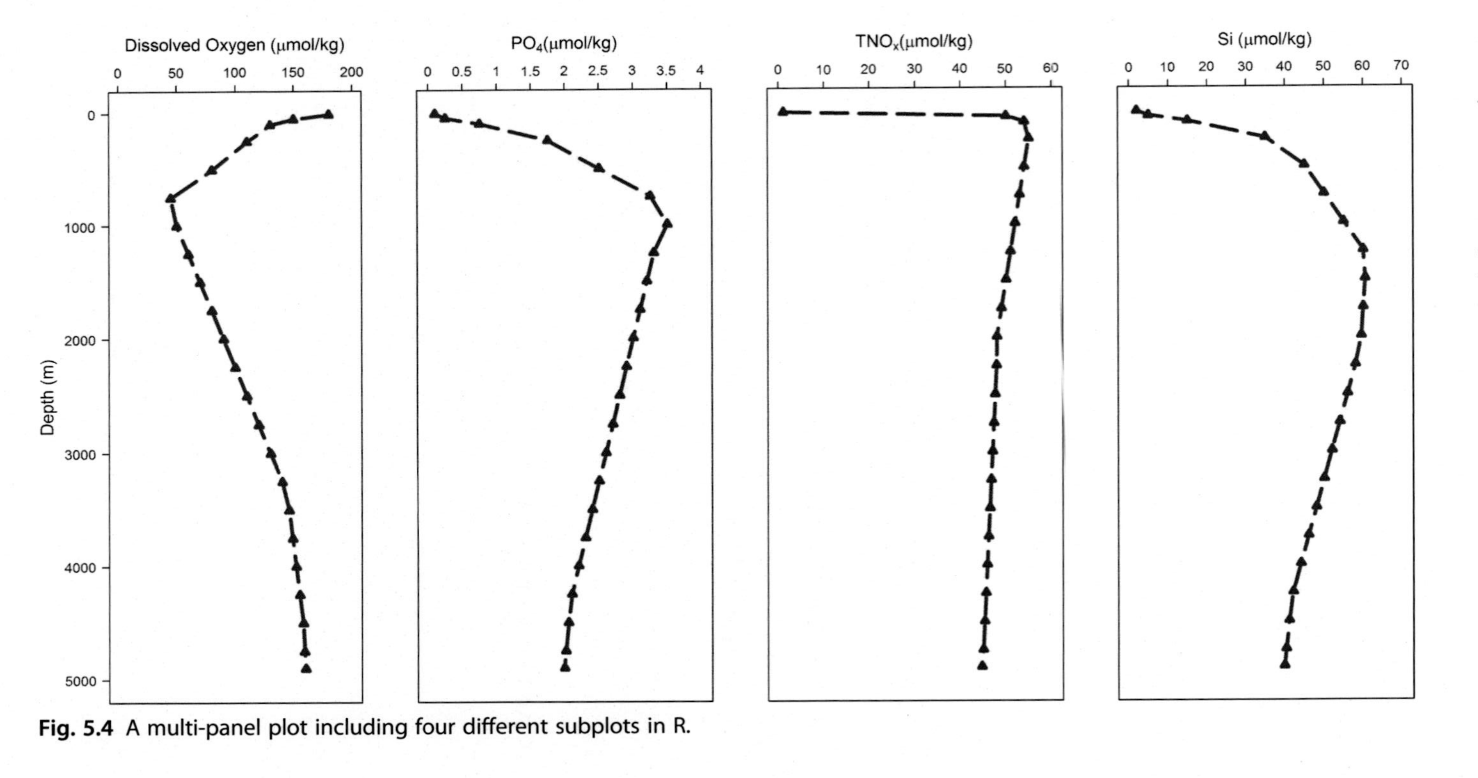

Fig. 5.4 A multi-panel plot including four different subplots in R.

Define the plot
Note that before each plot, par function is used to create new subplots
Subplot 1

```
plot(0, 0, type="o", xaxt="n", yaxt="n", xlim=c(0,200),
     ylim=rev(range(0,5000)), col="white", pch=0, cex=1,
     lwd=2, main="",  xlab="", ylab="")
axis(2, at=y_ticks, labels=y_ticks, cex.axis=1)
axis(3, at=x1_ticks, labels=x1_ticks, cex.axis=1)
mtext(expression("Dissolved  Oxygen  ("*mu*"mol/kg)"),  side=3,
line=2.6, cex=1, las=1, col="black")
mtext("Depth (m)", side=2, line=3.5, cex=1, las=0,
col="black")
lines(data$do,data$depth, col="blue", pch=2, type="o",
lty="longdash", lwd=3)
```

Subplot 2

```
par(mar=c(1,3,4,1))
plot(0, 0, type="o", xaxt="n", yaxt="n", xlim=c(0,4),
     ylim=rev(range(0,5000)), col="white", pch=0, cex=1,
     lwd=2, main="",  xlab="", ylab="")
axis(3, at=x2_ticks, labels=x2_ticks, cex.axis=1)
mtext(expression("PO"[4]*"("*mu*"mol/kg)"), side=3,
line=2.6, cex=1, las=1, col="black")
lines(data$phos, data$depth, col="blue", pch=2, type="o",
lty="longdash", lwd=3)
```

Subplot 3

```
par(mar=c(1,3,4,1))
plot(0, 0, type="o", xaxt="n", yaxt="n", xlim=c(0,60),
     ylim=rev(range(0,5000)), col="white", pch=0, cex=1,
     lwd=2, main="",  xlab="", ylab="")
axis(3, at=x3_ticks, labels=x3_ticks, cex.axis=1)
mtext(expression("TNO"[x]*"("*mu*"mol/kg)"), side=3,
line=2.6, cex=1, las=1, col="black")
lines(data$tnox, data$depth, col="blue", pch=2, type="o",
lty="longdash", lwd=3)
```

Subplot 4

```
par(mar=c(1,3,4,1))
plot(0, 0, type="o", xaxt="n", yaxt="n", xlim=c(0,70),
     ylim=rev(range(0,5000)), col="white", pch=0, cex=1,
     lwd=2, main="",  xlab="", ylab="")
axis(3, at=x4_ticks,labels=x4_ticks, cex.axis=1)
mtext(expression("Si ("*mu*"mol/kg)"), side=3, line=2.6,
cex=1, las=1, col="black")
lines(data$si,data$depth, col="blue", pch=2, type="o",
lty="longdash", lwd=3)
dev.off()
```

5.2 Time-series plots in R

The following dataset will be used to produce time-series plots. There are 64 observations in the dataset. There are three variables: date, density, and salinity. Values do not belong to field observations.

	date	dens	salin
1	2003-09-19	25.92	36.42
2	2003-09-18	28.82	36.90
3	2003-09-17	29.43	38.33
4	2003-09-16	25.84	36.81
5	2003-09-15	28.08	39.51
6	2003-09-12	29.25	37.64
7	2003-09-11	27.93	35.39
8	2003-09-10	29.69	38.21
9	2003-09-09	26.26	38.31
10	2003-09-08	29.69	38.78
11	2003-09-05	29.22	38.70
12	2003-09-04	29.74	35.16
13	2003-09-03	29.94	34.85
14	2003-09-02	25.22	37.60
15	2003-08-29	27.62	37.38
16	2003-08-28	26.22	37.58
17	2003-08-27	26.47	38.25
18	2003-08-26	28.93	38.42
19	2003-08-25	28.62	37.44
20	2003-08-22	29.86	36.99
21	2003-08-21	28.32	39.96

```
22   2003-08-20   29.48   37.57
23   2003-08-19   24.96   35.11
24   2003-08-18   27.40   37.15
25   2003-08-15   26.86   37.24
26   2003-08-14   27.95   39.49
27   2003-08-13   25.38   37.96
28   2003-08-12   24.54   38.28
29   2003-08-11   25.04   34.43
30   2003-08-08   25.51   37.52
31   2003-08-07   25.78   36.26
32   2003-08-06   28.88   35.42
33   2003-08-05   27.34   39.68
34   2003-08-04   26.46   37.10
35   2003-08-01   27.49   37.71
36   2003-07-31   28.82   39.25
37   2003-07-30   28.73   38.08
38   2003-07-29   25.70   37.76
39   2003-07-28   26.76   35.85
40   2003-07-25   26.95   37.15
41   2003-07-24   29.86   33.80
42   2003-07-23   26.91   38.15
43   2003-07-22   27.59   36.38
44   2003-07-21   29.76   37.29
45   2003-07-18   28.02   36.65
46   2003-07-17   27.81   38.41
47   2003-07-16   29.36   37.10
48   2003-07-15   29.54   39.36
49   2003-07-14   28.42   37.93
50   2003-07-11   27.78   36.45
51   2003-07-10   26.94   40.15
52   2003-07-09   25.89   36.15
53   2003-07-08   29.25   37.37
54   2003-07-07   27.99   39.50
55   2003-07-03   26.27   34.05
56   2003-07-02   27.22   36.89
57   2003-07-01   28.20   36.32
58   2003-06-30   28.77   39.11
59   2003-06-27   28.67   35.73
60   2003-06-26   25.80   39.06
61   2003-06-25   28.97   35.36
62   2003-06-24   24.92   41.03
63   2003-06-23   25.75   36.35
64   2003-06-20   25.48   37.13
```

A time-series plot for a single variable (Fig. 5.5):

Define working folder path and load data file

```
setwd("/path-to-local-data-folder/")
data <- read.table("data.txt", header=TRUE)
attach(data)
```

Introduce date values in ISO format

```
data$date = as.Date(data$date, format='%Y-%m-%d')
```

Define y axis ticks manually with seq function

```
y1list <- seq(floor(min(data$dens))-1,
ceiling(max(data$dens))+1, 1)
y2list <- seq(floor(min(data$salin)),
ceiling(max(data$salin)), 1)
```

Define x axis tick positions

```
xaxpos <- c(data$date[6], data$date[15], data$date[25],
data$date[35], data$date[45], data$date[54], data$date[64])
```

Define output folder path for figure and set figure parameters

```
tiff("figure-name.tiff",   width=28,   height=12,   unit="cm",
res=600, pointsize=10, compression="lzw", bg="white")
par(mar=c(2,4.5,2,2), las=1, xpd=TRUE)
```

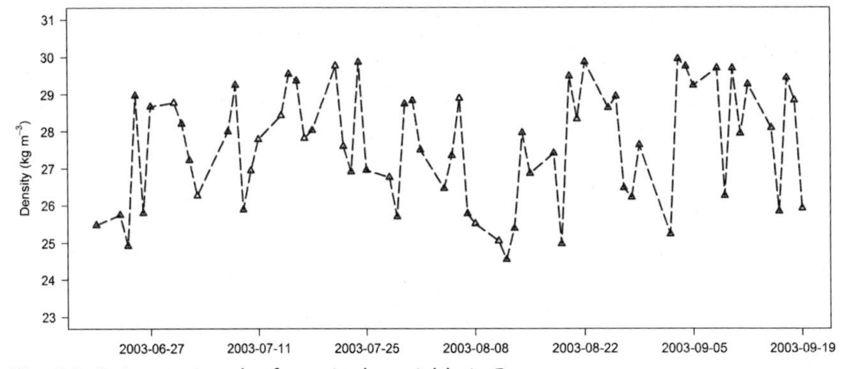

Fig. 5.5 A time-series plot for a single variable in R.

Define the plot
Define and format x axis with axis.Date() function

```
plot(data$date,data$dens,type="o", xaxt="n", yaxt="n",
     axes=FALSE,  col="white", pch=0, cex=1, lwd=2, main="",
     xlab="", ylab="",  cex.axis=1.2, cex.lab=1.2,
     ylim=c(floor(min(data$dens))-1,
     ceiling(max(data$dens))+1))
axis.Date(1, at = xaxpos, format= "%Y-%m-%d", las = 1),
format= "%Y-%m-%d",  las = 1)
axis(2, at=yllist, labels=yllist, cex.axis=1)
lines(data$date,data$dens, col="blue", pch=2, type="o",
lty="longdash", lwd=2)
mtext(2, text=expression("Density (kg m"^-3*")"), line=2.5,
cex=1, las=0)
box()
dev.off()
```

A time-series plot for two variables given on two different y-axis (Fig. 5.6):
Define working folder path and load data file

```
setwd("/path-to-local-data-folder/")
data <- read.table("date.txt", header=TRUE)
attach(data)
```

Introduce date values in ISO format

```
data$date = as.Date(data$date, format='%Y-%m-%d')
```

Define y axis ticks manually with seq function

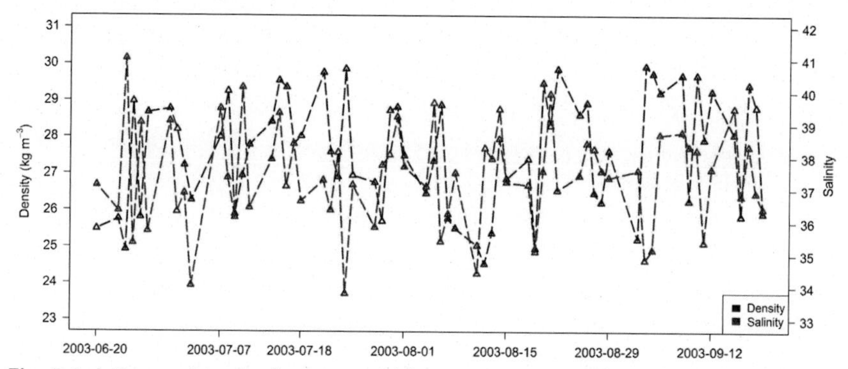

Fig. 5.6 A time-series plot for two variables given on two different y-axis in R.

```
y1list <- seq(floor(min(data$dens))-1,
ceiling(max(data$dens))+1,1)
y2list <-
seq(floor(min(data$salin)), ceiling(max(data$salin)),1)
```

Define x axis tick positions

```
xaxpos <- c(data$date[6], data$date[15], data$date[25],
data$date[35], data$date[45], data$date[54], data$date[64])
```

Define output folder path for figure and set figure parameters

```
tiff("figure-name.tiff", width=28, height=12, unit="cm",
res=600, pointsize=10, compression="lzw", bg="white")
par(mar=c(2,4.5,2,4.5), las=1, xpd=TRUE)
```

Define the plot

```
plot(data$date, data$dens, type="o", xaxt="n", yaxt="n",
     axes=FALSE,  col="white", pch=0, cex=1, lwd=2, main="",
     xlab="", ylab="",  cex.axis=1.2, cex.lab=1.2,
     ylim=c(floor(min(data$dens))-1,
     ceiling(max(data$dens))+1))
axis.Date(1, at = xaxpos, format= "%Y-%m-%d", las = 1)
axis(2, at=y1list, labels=y1list, cex.axis=1)
lines(data$date,data$dens, col="blue", pch=2, type="o",
lty="longdash", lwd=2)
mtext(2, text=expression("Density (kg m"^-3*")"), line=2.5,
cex=1, las=0)
par(new=T)
plot(data$date, data$salin, col="red", pch=2, type="o",
     lty="longdash",  lwd=2, xaxt="n", yaxt="n", axes=FALSE,
     main="", xlab="", ylab="",
     ylim=c(floor(min(data$salin)),
     ceiling(max(data$salin))))
axis(4, at=y2list, labels=y2list, cex.axis=1)
mtext(4, text=expression("Salinity"), line=2.5, cex=1,
las=0)
box()
dev.off()
```

5.3 Barplots in R

The following dataset (can be saved as "grouped_data.txt") will be used to produce barplots, boxplots, pie charts, and 3d scatter plots in this chapter. There are 23 observations in the dataset and vertical Dissolved Oxygen (do), ortho-Phosphate (phos), Total Nitrite + Nitrate (tnox), Silicate (si), and calculation (calc) data is given. Dataset is grouped according to group, phos25, limit and range variables. Values do not belong to field observations.

group	phos25	limit	range	depth	do	phos	tnox	si	calc
A	pb25	bel2000	1000	10	180	0.1	1.0	2.0	-1.0
A	pb25	bel2000	1000	50	150	0.3	50.0	5.0	4.5
A	pb25	bel2000	1000	100	130	0.8	54.0	15.0	3.9
A	pb25	bel2000	1000	250	110	1.8	55.0	35.0	2.0
B	pa25	bel2000	1000	500	80	2.5	54.0	45.0	6.0
B	pa25	bel2000	1000	750	45	3.3	53.0	50.0	3.0
B	pa25	bel2000	1000	1000	50	3.5	52.0	55.0	4.0
B	pa25	bel2000	2000	1250	60	3.3	51.0	60.0	5.0
C	pa25	bel2000	2000	1500	70	3.2	50.0	60.5	1.5
C	pa25	bel2000	2000	1750	80	3.1	49.0	60.0	-1.0
C	pa25	bel2000	2000	2000	90	3.0	48.0	59.5	-1.5
C	pa25	ab2000	3000	2250	100	2.9	47.9	58.0	-1.0
D	pa25	ab2000	3000	2500	110	2.8	47.6	56.0	-6.4
D	pa25	ab2000	3000	2750	120	2.7	47.3	54.0	-6.7
D	pa25	ab2000	3000	3000	130	2.6	47.0	52.0	-5.0
D	pa25	ab2000	4000	3250	140	2.5	46.7	50.0	-3.3
E	pb25	ab2000	4000	3500	146	2.4	46.4	48.0	-1.6
E	pb25	ab2000	4000	3750	149	2.3	46.1	46.0	0.1
E	pb25	ab2000	4000	4000	152	2.2	45.8	44.0	1.8
E	pb25	ab2000	5000	4250	155	2.1	45.5	42.0	3.5
E	pb25	ab2000	5000	4500	158	2.1	45.2	41.0	4.2
E	pb25	ab2000	5000	4750	159	2.0	44.9	40.2	4.7
E	pb25	ab2000	5000	4900	160	2.0	44.6	39.8	4.8

A barplot for a single dependent variable versus a factor variable (Fig. 5.7):

Define working folder path and load data file

```
setwd("/path-to-local-data-folder/")
data <- read.table("grouped_data.txt", header=TRUE)
attach(data)
```

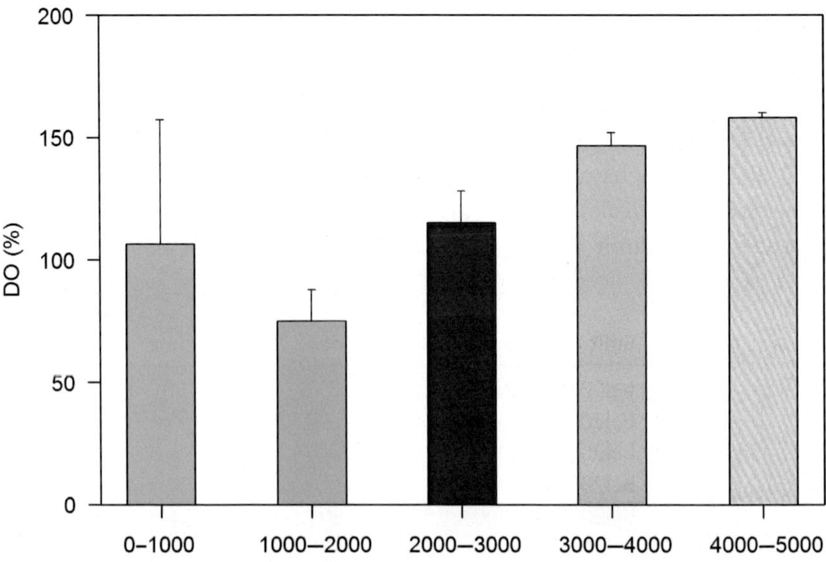

Fig. 5.7 A barplot for a single dependent variable versus a factor variable in R.

Define dependent and factor variables; calculate means with errorbars

```
data$range <- factor(data$range,
levels=c("1000","2000","3000","4000","5000"))

dfagg <- aggregate(data$do, by=list(range=data$range),
FUN=function(x) c(mean=mean(x), sd=sd(x), n=length(x)))

newdf <- do.call(data.frame, dfagg)

colnames(newdf ) <- c("range", "mean", "sd", "n")

do_mean <- tapply(newdf$mean, list(newdf$range), function(x)
c(x=x))

do_sd <- tapply(newdf$sd, list(newdf$range), function(x)
c(x=x))
```

Define the plot

```
tiff(filename="figure-name.tiff", width=14,height=10,
unit="cm", res=300, pointsize=10, compression="lzw",
bg="white")
```

```
par(mar=c(4,4,1.5,1.5), las=1)

bars <- barplot(height=do_mean, beside=TRUE, las=1,
     cex.names=1, main="", ylim=c(0,200), ylab="DO (%)",
     xlab="", yaxs="i", width=0.8, space=1.2,
     border="black", axes=TRUE, legend=FALSE, axis.lty= 1,
     col=c("lightskyblue","greenyellow","firebrick1","gold",
     "gray88"), names.arg=c("0-1000","1000-2000","2000-
     3000","3000-4000","4000-5000"))
segments(bars, do_mean, bars, do_mean + do_sd, lwd=0.7)
arrows(bars, do_mean, bars, do_mean + do_sd, lwd=0.7,
angle=90, code=3, length=0.025)
box()
dev.off()
```

A barplot for a single dependent variable versus two factor variables (Fig. 5.8):
Define working folder path and load data file

```
setwd("/path-to-local-data-folder/")
data <- read.table("grouped_data.txt", header=TRUE)
attach(data)
```

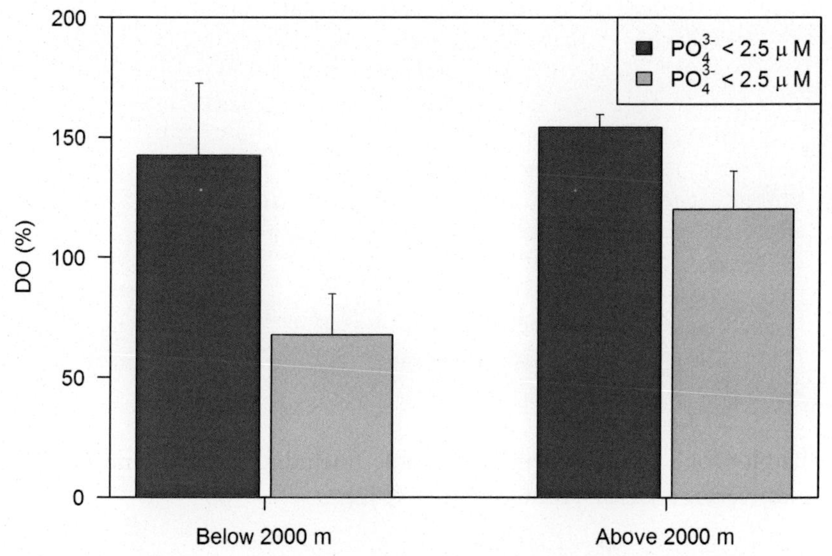

Fig. 5.8 A barplot for a single dependent variable versus two factor variables in R.

Define dependent and factor variables; calculate means with errorbars

```
df <- subset(data, select=c(limit,phos25,do))

df$limit <- factor(df$limit, levels=c("bel2000","ab2000"))
df$phos25 <- factor(df$phos25, levels=c("pb25","pa25"))

dfagg <- aggregate(df$do, by=list(phos25=df$phos25,
limit=df$limit), FUN=function(x) c(mean=mean(x), sd=sd(x),
n=length(x)))

newdf <- do.call(data.frame, dfagg)

colnames(newdf ) <- c("limit", "phos25", "mean", "sd", "n")

do_mean <- tapply(newdf$mean, list(newdf$limit,
newdf$phos25), function(x) c(x=x))

do_sd  <-  tapply(newdf$sd,  list(newdf$limit,  newdf$phos25),
function(x) c(x=x))
```

Define the plot

```
tiff(filename="figure-name.tiff", width=14,height=10,
unit="cm", res=300, pointsize=10, compression="lzw",
bg="white")
par(mar=c(4,4,1.5,1.5), las=1)

bars <- barplot(height=do_mean, beside=TRUE, las=1,
      cex.names=1, main="", ylim=c(0,200), ylab="DO (%)",
      xlab="", yaxs="i", space=c(0.6,0.1,1.2,0.1),
      border="black", axes=TRUE, legend=FALSE, axis.lty=1,
      col=c("orangered","gold"),
      names.arg=c("Below 2000 m","Above 2000 m"))
segments(bars, do_mean, bars, do_mean + do_sd, lwd=0.7)
arrows(bars, do_mean, bars, do_mean + do_sd, lwd=0.7,
angle=90, code=3, length=0.025)
box()
legend("topright", legend=c("PO4 < 2.5 uM","PO4 > 2.5 uM"),
fill=c("orangered","gold"), horiz=FALSE)
dev.off()
```

A barplot for a single dependent variable (including positive and negative values) versus a factor variable (Fig. 5.9) [2]:

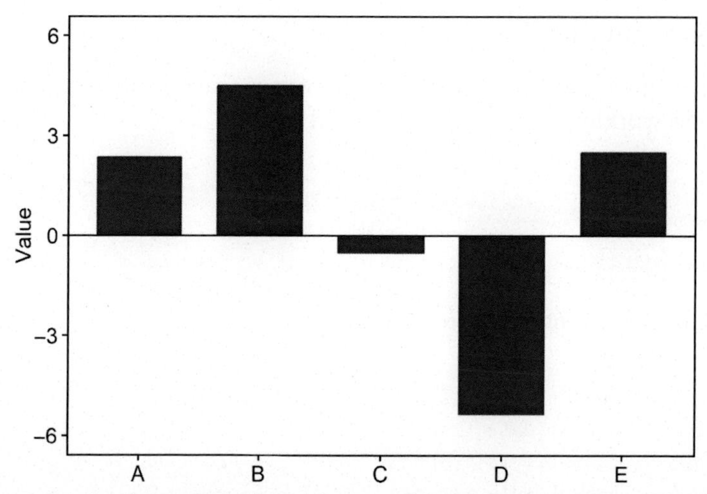

Fig. 5.9 A barplot for a single dependent variable (including positive and negative values) versus a factor variable in R.

Define working folder path and load data file

```
setwd("/path-to-local-data-folder/")
data <- read.table("grouped_data.txt", header=TRUE)
attach(data)
```

Define dependent and factor variables

```
library(ggpubr)

df <- data.frame(group=data$group, calc=data$calc)
df2 <- data.frame(x=c(1,2,3,4,5), y=c(0,0,0,0,0))
```

Define the plot

```
tiff(filename="figure-name.tiff", width=14,height=10,
unit="cm", res=300, pointsize=10, compression="lzw",
bg="white")
par(mar=c(4,4,1.5,1.5), las=1)

p <- ggbarplot(df, x = "group", y = "calc", add="mean",
     color = "darkblue", fill="darkblue", ylab = "Value",
     xlab="") + border() + geom_exec(geom_hline, yintercept
     = 0)
ggpar(p, ylim = c(-6,6))
```

5.4 Boxplots in R

A boxplot for a single dependent variable versus a factor variable (Fig. 5.10):
Define working folder path and load data file

```
setwd("/path-to-local-data-folder/")
data <- read.table("grouped_data.txt", header=TRUE)
attach(data)
```

Define dependent and factor variables

```
range1 <- subset(data, range=="1000", select=c(do))
range2 <- subset(data, range=="2000", select=c(do))
range3 <- subset(data, range=="3000", select=c(do))
range4 <- subset(data, range=="4000", select=c(do))
range5 <- subset(data, range=="5000", select=c(do))

q <- quantile(data$do, names=FALSE)
```

Define the plot

```
tiff(filename="figure-name.tiff", width=18,height=12,
unit="cm", res=300, pointsize=12, compression="lzw",
```

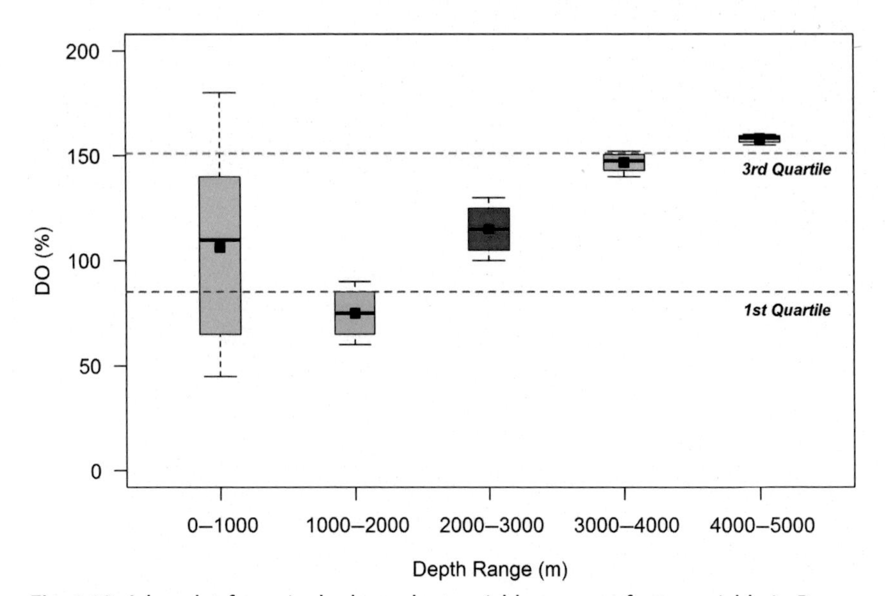

Fig. 5.10 A boxplot for a single dependent variable versus a factor variable in R.

```
bg="white")
par(mar=c(4,4,1.5,1.5), las=1)
boxplot(data$do ~ data$range,
     at = c(1,2,3,4,5),
     names=c("0-1000","1000-2000","2000-3000","3000-
     4000","4000-5000"),
     ylim=c(0,200), main = "", xlab="", ylab="DO (%)",
     pars=list(boxwex=0.3, staplewex=0.8, outwex=0.8),
     col=c("darkslategray1","greenyellow","orangered",
     "gold","gray88"))
mtext("Depth Range (m)",side=1,line=2.75,cex=1)
points(1,mean(range1$do),pch=22,bg="black",cex=1.2)
points(2,mean(range2$do),pch=22,bg="black",cex=1.2)
points(3,mean(range3$do),pch=22,bg="black",cex=1.2)
points(4,mean(range4$do),pch=22,bg="black",cex=1.2)
points(5,mean(range5$do),pch=22,bg="black",cex=1.2)
text(5.2, 0.90*q[2], expression(bolditalic("1st
Quartile")),cex=0.8)
text(5.2, 0.95*q[4], expression(bolditalic("3rd
Quartile")),cex=0.8)
abline(h=q[2], lty=2, lwd=1.5, col="red")
abline(h=q[4], lty=2, lwd=1.5, col="darkgreen")
dev.off()
```

A boxplot for a single dependent variable versus two factor variables (Fig. 5.11):
Define working folder path and load data file

```
setwd("/path-to-local-data-folder/")
data <- read.table("grouped_data.txt", header=TRUE)
attach(data)
```

Define dependent and factor variables

```
data$limit <- factor(data$limit,
levels=c("bel2000","ab2000"))
data$phos25 <- factor(data$phos25, levels=c("pb25","pa25"))
```

Define the plot

```
tiff(filename="figure-name.tiff", width=18,height=15,
unit="cm", res=300, pointsize=12, compression="lzw",
bg="white")
par(mar=c(4,4,1.5,1.5), las=1)

a <- boxplot(data$do ~ data$phos25 * data$limit,
     at = c(1:2,4:5),
     ylim=c(0,200),
     main = "",
     xlab="",
     ylab="DO (%)",
     pars=list(boxwex=0.6, staplewex=0.8, outwex=0.8),
     col=c("orangered","gold"),
     xaxs=FALSE, xaxt="n")
axis(1,at=c(1.5,4.5),
     labels=c("Below 2000 m","Above 2000 m"))
legend("topright", legend=c("PO4 < 2.5 uM","PO4 > 2.5 uM"),
fill=c("orangered","gold"), horiz=FALSE)
dev.off()
```

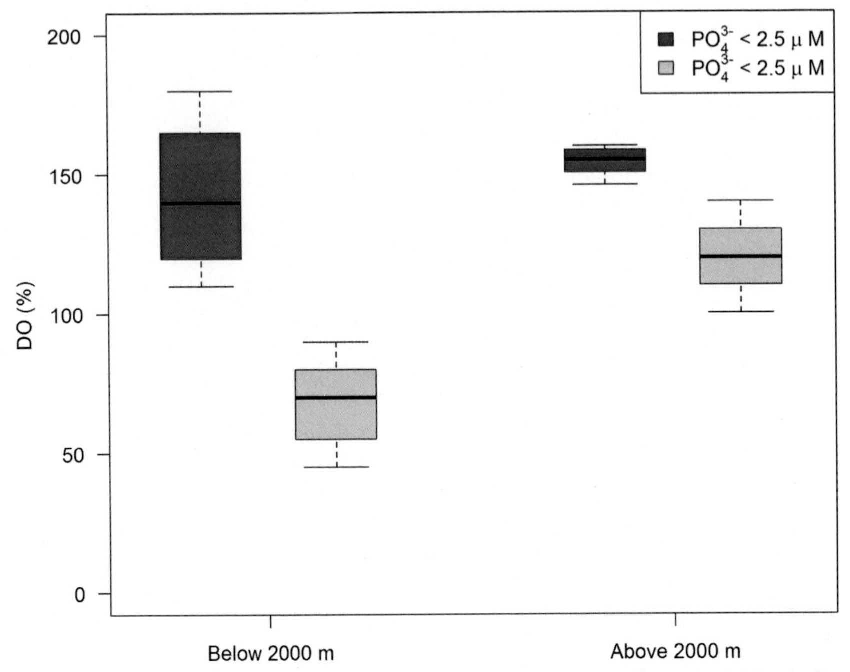

Fig. 5.11 A boxplot for a single dependent variable versus two factor variables in R.

5.5 Pie charts in R

A 2d pie chart (Fig. 5.12):
Define working folder path and load data file

```
setwd("/path-to-local-data-folder/")
data <- read.table("grouped_data.txt", header=TRUE)
attach(data)
```

Define dependent and factor variables

```
data$group <- factor(data$group,
levels=c("A","B","C","D","E"))

dfagg <- aggregate(data$si, by=list(group=data$group),
FUN=function(x) c(mean=mean(x), sd=sd(x), n=length(x)))

newdf <- do.call(data.frame, dfagg)

colnames(newdf ) <- c("group", "mean", "sd", "n")

si_mean <- tapply(newdf$mean, list(newdf$group), function(x)
c(x=x))

s = list()

for (i in 1:length(si_mean)){
  s[i] = si_mean[i]/sum(si_mean)*100
  output = cbind(s)
}

s = c(s[[1]],s[[2]],s[[3]],s[[4]],s[[5]])

lbls <- c("A:", "B:", "C:", "D:", "E:")
pie_lbl <- paste(lbls, round(s, digits=0)) # add percents to
labels
pie_lbl <- paste(pie_lbl,"%",sep="") # ad % to labels
```

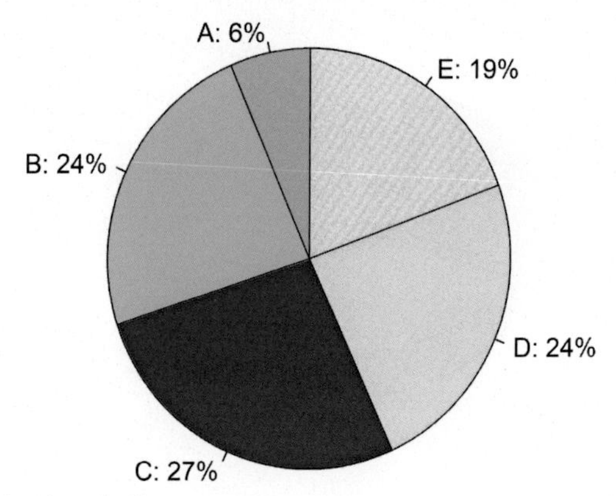

Fig. 5.12 A 2d pie chart in R.

Define the plot

```
tiff(filename="figure-name.tiff", width=10,height=10,
unit="cm", res=600, pointsize=12, compression="lzw",
bg="white")
par(mfrow=c(1,1), mar=c(1,2.5,3,3), las=1)

pie(s, labels=pie_lbl, main="", init.angle=90,
     col=c("chartreuse","skyblue2","red","yellow","grey90"),
     edges=400, radius=1.0)
dev.off()
```

A 3d pie chart plotted with plotrix package (Fig. 5.13) [3]:
Define working folder path and load data file
Define working folder

```
setwd("/path-to-local-data-folder/")
data <- read.table("grouped_data.txt", header=TRUE)
attach(data)
```

Define dependent and factor variables

```
library(plotrix)

data$group <- factor(data$group,
levels=c("A","B","C","D","E"))

dfagg <- aggregate(data$si, by=list(group=data$group),
FUN=function(x) c(mean=mean(x), sd=sd(x), n=length(x)))
```

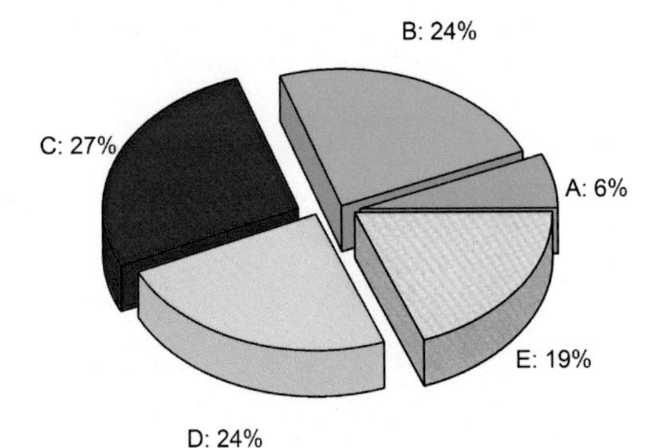

Fig. 5.13 A 3d pie chart in R.

```
newdf <- do.call(data.frame, dfagg)

colnames(newdf ) <- c("group", "mean", "sd", "n")

si_mean <- tapply(newdf$mean, list(newdf$group), function(x)
c(x=x))

s = list()

for (i in 1:length(si_mean)){
  s[i] = si_mean[i]/sum(si_mean)*100
  output = cbind(s)
}

s = c(s[[1]],s[[2]],s[[3]],s[[4]],s[[5]])

lbls <- c("A:", "B:", "C:", "D:", "E:")
pie_lbl <- paste(lbls, round(s, digits=0)) # add percents to
labels
pie_lbl <- paste(pie_lbl,"%",sep="") # ad % to labels
```

Define the plot

```
tiff(filename="figure-name.tiff", width=10,height=10,
unit="cm", res=600, pointsize=12, compression="lzw",
bg="white")
par(mfrow=c(1,1), mar=c(1,2.5,3,3), las=1)

pie3D(s, labels=pie_lbl, labelcex=0.7, main="",
     explode=0.15, start=0, theta=pi/3,
     col=c("chartreuse","skyblue2","red","yellow","grey90"),
     edges=1000, radius=0.9)
dev.off()
```

5.6 3D plots in R

A 3d scatter plot produced with OceanView package (Fig. 5.14) [4]:

Define working folder path and load data file

```
setwd("/path-to-local-data-folder/")
data <- read.table("grouped_data.txt", header=TRUE)
attach(data)
```

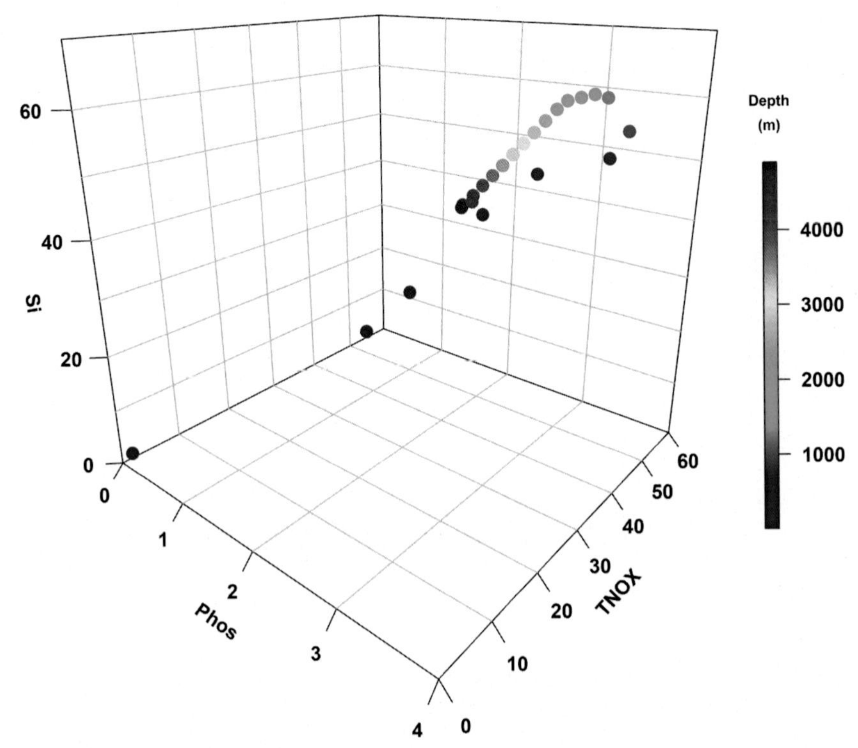

Fig. 5.14 A 3d scatter plot in R.

Define dependent and factor variables

```
library(psych)
library(OceanView)
df <- subset(data, select=c(phos,tnox,si,depth))
x <- df$phos
y <- df$tnox
z <- df$si
cv <- df$depth
names <- df$depth
```

Define the plot

```
tiff(filename="figure-name.tiff", width=20,height=20,
unit="cm", res=600, pointsize=12, compression="lzw",
bg="white")
par(mfcol=c(1,1),  las=1,  font=2,  font.axis=2,  font.lab=2,
cex=1, cex.axis=1, cex.lab=1)
scatter3D(x=x, y=y, z=z, colvar=cv,
     pch=16, cex=1.5, font=2,
```

```
      ticktype="detailed", col=jet.col(250), phi=20,
      theta=40, bty="b2",
      xlab="Phos", ylab="TNOX", zlab="Si",
      clab=c("Depth","(m)"),
      xlim=c(0,4),xaxp=c(0,4,4),
      ylim=c(0,60),yaxp=c(0,60,6),
      zlim=c(0,70),zaxp=c(0,70,4),
      colkey = list(length = 0.5, width = 0.5, cex.clab = 0.75))
dev.off()
```

A 3d surface plot produced with plot3D package (Fig. 5.15) [5]:

Define working folder path and load data file
Define working folder

```
setwd("/path-to-local-data-folder/")
```

Define variables

```
library(plot3D)

X <- seq(0, 2*pi, length.out = 100)
Y <- seq(pi/12, pi/2, length.out = 100)

M       <- mesh(X, Y)
phi     <- M$x
theta   <- M$y
# x, y and z grids
r <- sin(2*phi)+cos(3*theta)
x <- phi/3
y <- theta/3
z <- r
```

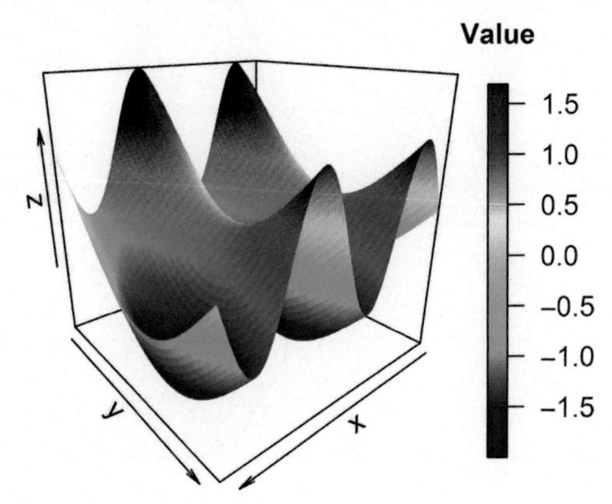

Fig. 5.15 A 3d surface plot in R.

Define the plot

```
tiff(filename="figure-name.tiff", width=10,height=10,
unit="cm", res=600, pointsize=12, compression="lzw",
bg="white")
par(mfrow=c(1,1), mar=c(1,2.5,3,2), las=1)

surf3D(x, y, z, colvar = z, bty="b", shade = 0.1,
     colkey = list(plot=TRUE, length=0.7), box = TRUE,
     phi=20, theta =140, lighting = TRUE, clab="Value")
dev.off()
```

5.7 Ternary plots in R

Ternary diagrams can be plotted by using ggtern package (Fig. 5.16) [6]:
Define working folder path and load data file

```
setwd("/path-to-local-data-folder/")
data <- read.table("grouped_data.txt", header=TRUE)
attach(data)
```

Define variables

```
library(ggtern)

x <-
data.frame(a=data[,7],b=data[,8],c=data[,9],k=data[,1])
```

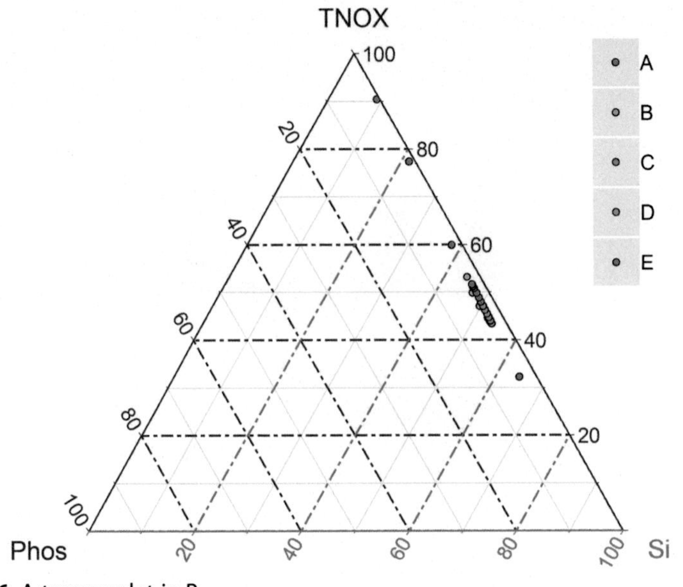

Fig. 5.16 A ternary plot in R.

Define the plot

```
tiff("figure-name.tiff", width=16,height=12, unit="cm",
res=600, pointsize=16, compression="lzw", bg="white")
par(mar=c(0,0,0,0), las=1, xpd=TRUE)

ggtern(data=x,aes(a,b,c))
     + theme_bw(base_size=8)
     + custom_percent('%')
     + geom_point(aes(fill=k),shape=21,size=1.25,alpha=1)
     + labs(title = "", x = "Phos", y = "TNOX", z = "Si")
     + theme_latex(TRUE)
     + theme_rgbw()
     + theme(legend.position = c(0.8,0.9),
          legend.justification = c(0,1),
          legend.title=element_blank())
     + theme_hidearrows()
     + guides(fill = guide_legend(order=1))
dev.off()
```

5.8 Vertical profiling plots in Python

A vertical profiling plot for a single variable (Fig. 5.17) [7]:

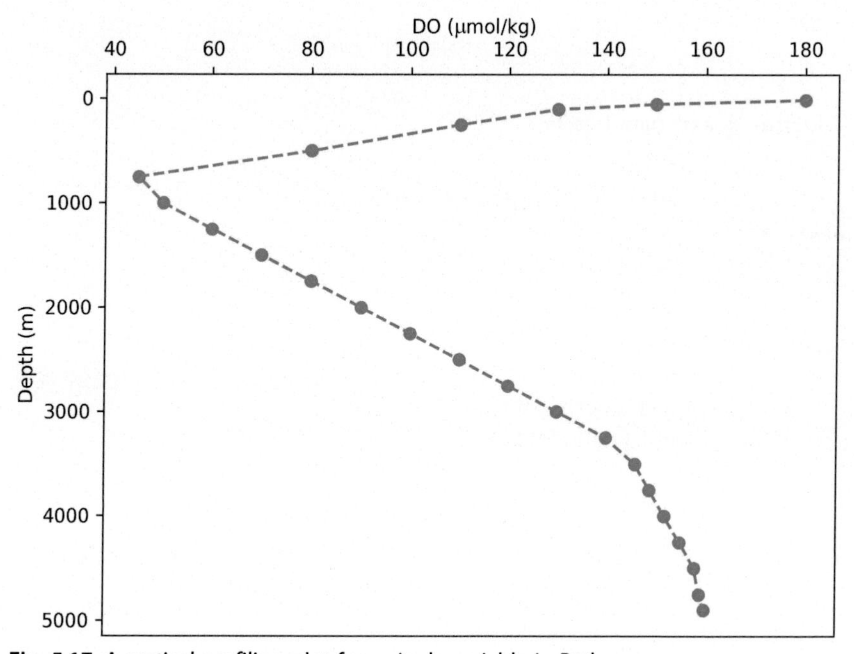

Fig. 5.17 A vertical profiling plot for a single variable in Python.

Load required libraries

```
import numpy as np
import matplotlib.pyplot as plt
```

Load data

```
f = open('/path-to-local-data-folder/data.txt', 'r')
data = np.genfromtxt(f, dtype="float", delimiter='\t',
names=True)
f.close()
```

Define variables

```
depth = data['depth']
do= data['do']
phos= data['phos']
del(data)
```

Define the plot

```
fig1 = plt.figure()
ax1 = fig1.add_subplot(111)
ax1.plot(do, depth, 'o--')
```

Define x-axis parameters

```
ax1.set_xlabel(r'DO ($\mu$mol/kg)')
ax1.xaxis.set_label_position('top')
ax1.xaxis.set_ticks_position('top')
```

Draw y-axis parameters and reverse y-axis

```
ax1.set_ylabel('Depth (m)')
ax1.set_ylim(ax1.get_ylim()[::-1])
```

Save and show the plot

```
plt.tight_layout()
plt.savefig('/path-to-figure-folder/name.png', format='png',
dpi=300, transparent=False)
plt.show()
```

A vertical profiling plot for two different variables plotted on the same axis (Fig. 5.18):
Load required libraries

```
import numpy as np
import matplotlib.pyplot as plt
```

Load data

```
f = open('/path-to-local-data-folder/data.txt', 'r')
data = np.genfromtxt(f, dtype="float", delimiter='\t',
names=True)
f.close()
```

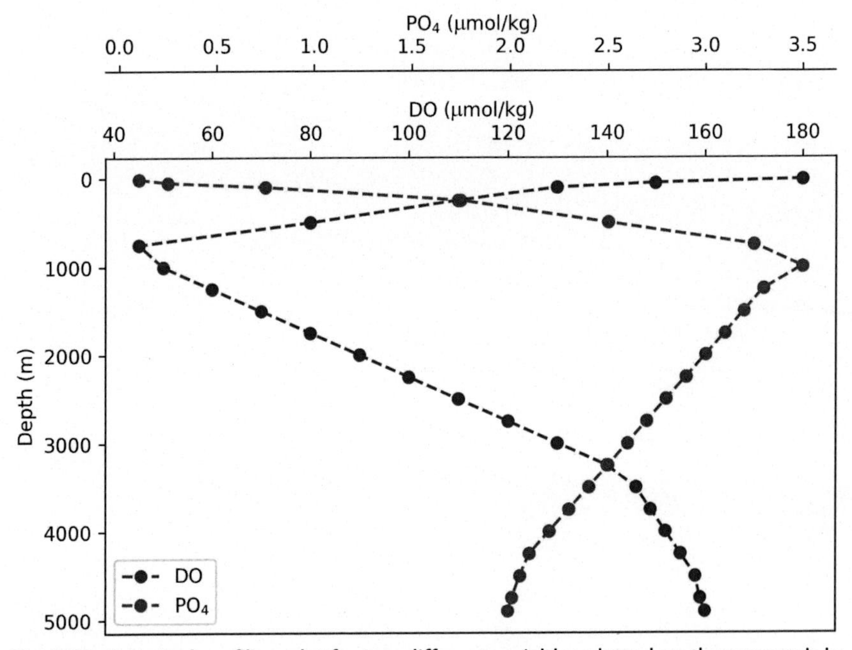

Fig. 5.18 A vertical profiling plot for two different variables plotted on the same axis in Python.

Define variables

```
depth = data['depth']
do= data['do']
phos= data['phos']
del(data)
```

Define the plot

```
fig=plt.figure()
ax1=fig.add_subplot(111, label='1')
```

Define first x-axis

```
s1,=ax1.plot(do, depth, 'o--', color='blue', label='DO')
ax1.set_ylim(ax1.get_ylim()[::-1])
ax1.set_xlabel(r'DO ($\mu$mol/kg)', color='black')
ax1.set_ylabel('Depth (m)', color='black')
```

Define second x-axis

```
dax1=ax1.twiny()
s2,=dax1.plot(phos, depth, 'o--', color='red', label=r'PO$_4$')
dax1.set_ylim(dax1.get_ylim()[::-1])
dax1.set_xlim(dax1.get_xlim())
dax1.set_xlabel(r'PO$_4$ ($\mu$mol/kg)', color='black')
dax1.yaxis.set_visible(False)
dax1.spines['top'].set_position(('outward', 45))
dax1.tick_params(axis='x', colors='black')
dax1.xaxis.tick_top()
dax1.xaxis.set_label_position('top')
dax1.xaxis.set_ticks_position('top')
```

Define parameters of the first x-axis

```
ax1.set_ylim(ax1.get_ylim()[::-1])
ax1.set_xlim(ax1.get_xlim())
ax1.tick_params(axis='x', colors='black')
ax1.tick_params(axis='y', colors='black')
ax1.xaxis.tick_top()
ax1.xaxis.set_label_position('top')
```

Define the plot legend

```
lns = [s1, s2]
ax1.legend(handles=lns, loc='lower left')
```

Save and show the plot

```
plt.tight_layout()
plt.savefig('/path-to-figure-folder/name.png', format='png',
dpi=300, transparent=False)
plt.show()
```

A vertical profiling plot for two different variables given on opposite axes (Fig. 5.19):
Load required libraries

```
import numpy as np
import matplotlib.pyplot as plt
```

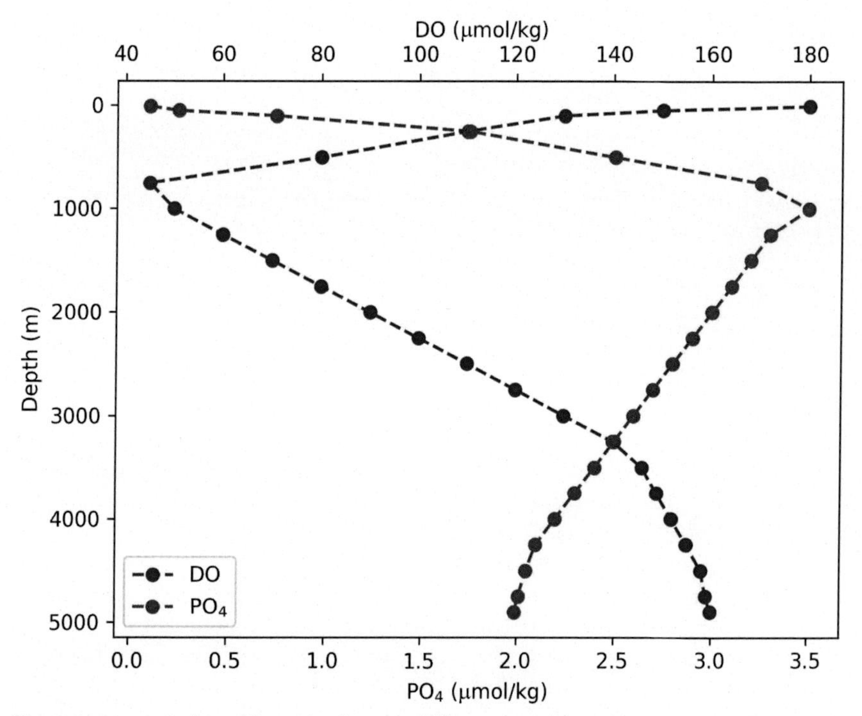

Fig. 5.19 A vertical profiling plot for two different variables given on opposite axes in Python.

Load data

```
f = open('/path-to-local-data-folder/data.txt', 'r')
data = np.genfromtxt(f, dtype="float", delimiter='\t',
names=True)
f.close()
```

Define variables

```
depth = data['depth']
do = data['do']
phos = data['phos']
del(data)
```

Define the plot

```
fig=plt.figure()
ax1=fig.add_subplot(111, label='1')
ax2=fig.add_subplot(111, label='2', frame_on=False)
```

Define first x-axis

```
s1,=ax1.plot(do, depth, 'o--', color='blue', label='DO')
ax1.set_ylim(ax1.get_ylim()[::-1])
ax1.set_xlabel(r'DO ($\mu$mol/kg)', color='black')
ax1.set_ylabel('Depth (m)', color='black')
ax1.tick_params(axis='x', colors='black')
ax1.tick_params(axis='y', colors='black')
ax1.xaxis.tick_top()
ax1.xaxis.set_label_position('top')
```

Define second x-axis

```
s2,=ax2.plot(phos, depth, 'o--', color='red', label=r'PO$_4$')
ax2.set_ylim(ax2.get_ylim()[::-1])
ax2.xaxis.tick_bottom()
ax2.set_xlabel(r'PO$_4$ ($\mu$mol/kg)', color='black')
ax2.xaxis.set_label_position('bottom')
ax2.tick_params(axis='x', colors='black')
ax2.yaxis.set_visible(False)
```

Define the plot legend

```
lns = [s1, s2]
ax1.legend(handles=lns, loc='lower left')
```

Save and show the plot

```
plt.tight_layout()
plt.savefig('/path-to-figure-folder/name.png', format='png',
dpi=300, transparent=False)
plt.show()
```

A multipanel plot including four different subplots (Fig. 5.20):
Load required libraries

```
import numpy as np
import matplotlib.pyplot as plt
```

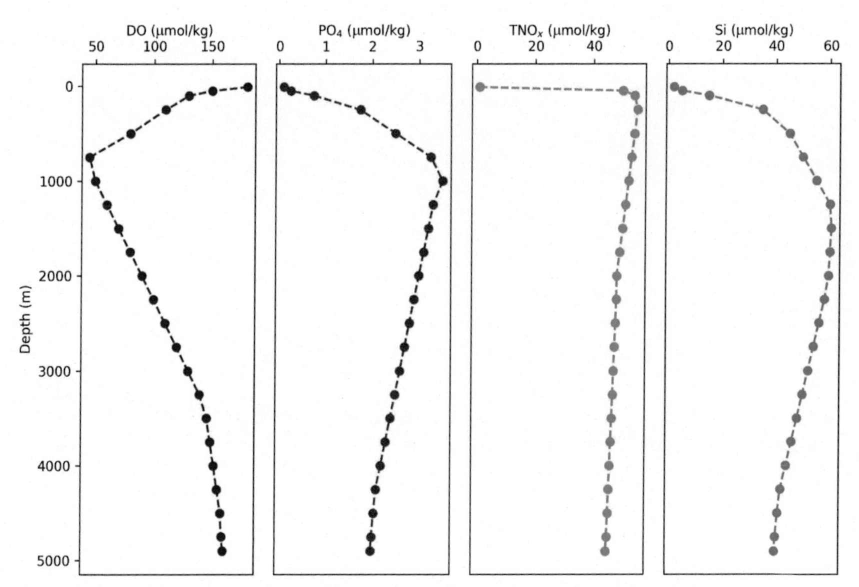

Fig. 5.20 A multi-panel plot including four different subplots in Python.

Load data

```
f = open('/path-to-local-data-folder/data.txt', 'r')
data = np.genfromtxt(f, dtype="float", delimiter='\t',
names=True)
f.close()
```

Define variables

```
depth = data['depth']
do = data['do']
phos = data['phos']
tnox = data['tnox']
si = data['si']
del(data)
```

Define the plot
Define subplot for do

```
fig1, (ax1, ax2, ax3, ax4) =
plt.subplots(1,4,sharey=True,figsize=(9, 6))
ax1.plot(do,depth,'o--b')
ax1.set_ylabel('Depth (m)')
ax1.set_ylim(ax1.get_ylim()[::-1])
ax1.set_xlabel(r'DO ($\mu$mol/kg)')
ax1.xaxis.set_label_position('top')
ax1.xaxis.set_ticks_position('top')
```

Define subplot for phosphate

```
ax2.plot(phos,depth,'o--r')
ax2.set_xlabel(r'PO$_4$ ($\mu$mol/kg)')
ax1.set_xlim(ax1.get_xlim())
ax2.xaxis.set_label_position('top')
ax2.xaxis.set_ticks_position('top')
ax2.yaxis.set_visible(False)
```

Define subplot for tnox

```
ax3.plot(tnox,depth,'o--g')
ax3.set_xlabel(r'TNO$_x$ ($\mu$mol/kg)')
ax3.xaxis.set_label_position('top')
ax3.xaxis.set_ticks_position('top')
ax3.yaxis.set_visible(False)
```

Define subplot for si

```
ax4.plot(si,depth,'o--')
ax4.set_xlabel(r'Si ($\mu$mol/kg)')
ax4.xaxis.set_label_position('top')
ax4.xaxis.set_ticks_position('top')
ax4.yaxis.set_visible(False)
```

Save and show the plot

```
plt.tight_layout()
plt.savefig('/path-to-figure-folder/name.png', format='png',
dpi=300, transparent=False)
plt.show()
```

5.9 Time-series plots in Python

A time-series plot for a single variable (Fig. 5.21) [7]:

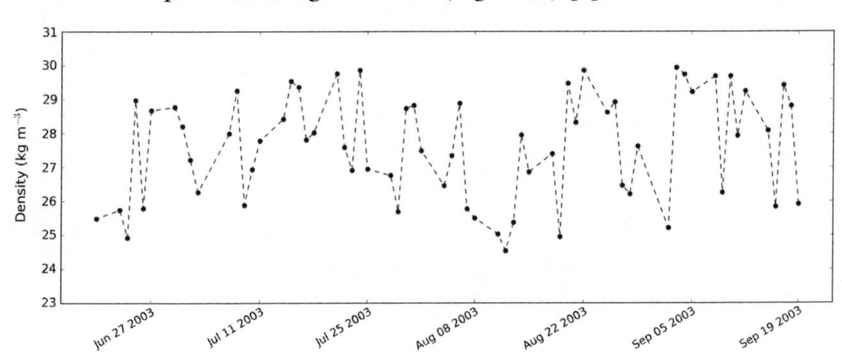

Fig. 5.21 A time-series plot for a single variable in Python.

Load required libraries

```
import datetime as dt
import numpy as np
import matplotlib.pyplot as plt
```

Define variables and load data

```
time=[]
dens=[]
dtime=[]

time = np.loadtxt('/path-to-local-data-folder/date.txt',
delimiter='\t', skiprows=1, usecols=[0], dtype='str')

dtime = [dt.datetime.strptime(d,'%Y-%m-%d') for d in time]

dens = np.loadtxt('/path-to-local-data-folder/date.txt',
delimiter='\t', skiprows=1, usecols=[1], dtype='float')
```

Define y axis limits

```
ymin = np.floor(np.min(dens)-1)
ymax = np.ceil(np.max(dens)+1)
```

Define the plot

```
fig = plt.figure(figsize=(15,6))
ax = plt.gca()
plt.plot_date(x=dtime, y=dens, marker='o', linestyle='dashed')
plt.xlabel('')
plt.ylabel('Density (kg m$^{-3}$)')
ax.set_ylim([ymin,ymax])
plt.margins(0.05, 0.1)
```

Format x-axis label as date

```
fig.autofmt_xdate()
```

Save and show the plot

```
plt.tight_layout()
plt.savefig('/path-to-figure-folder/name.png', format='png',
dpi=300, transparent=False)
plt.show()
```

A time-series plot for two variables given on two different y-axis (Fig. 5.22):
Load required libraries

```
import datetime as dt
import numpy as np
import matplotlib.pyplot as plt
from __future__ import unicode_literals
```

Define variables and load data

```
time=[]
dens=[]
salin=[]
dtime=[]

time = np.loadtxt('/path-to-local-data-folder/date.txt',
delimiter='\t', skiprows=1, usecols=[0], dtype='str')

dtime = [dt.datetime.strptime(d,'%Y-%m-%d') for d in time]

dens = np.loadtxt('/path-to-local-data-folder/date.txt',
delimiter='\t', skiprows=1, usecols=[1], dtype='float')

salin = np.loadtxt('/path-to-local-data-folder/date.txt',
delimiter='\t', skiprows=1, usecols=[2], dtype='float')
```

Define y axis limits

```
ymin = np.floor(np.min(dens)-1)
ymax = np.ceil(np.max(dens)+1)
```

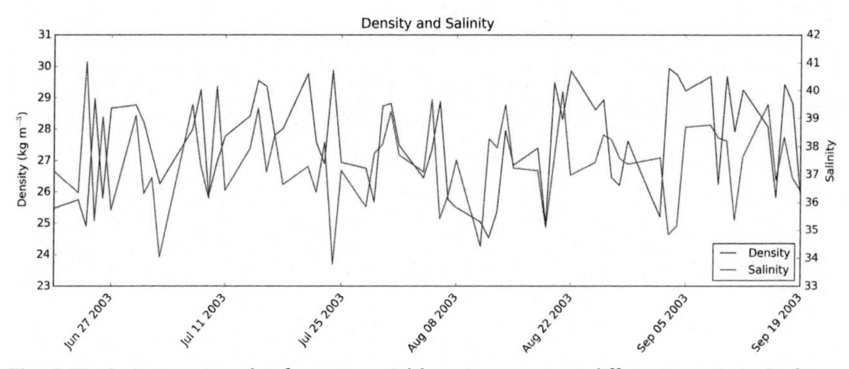

Fig. 5.22 A time-series plot for two variables given on two different y-axis in Python.

Define the plot

```
plt.rcParams['grid.linewidth'] = 0.5
plt.rcParams['grid.linestyle'] = '--'
fig, ax1 = plt.subplots(1, figsize=(15,6))
plt.xlabel('')
plt.ylabel(r'Density (kg m$^{-3}$)')
plt.tick_params(axis='y')
```

Define first y-axis

```
s1,=ax1.plot(dtime, dens, 'b-', linewidth=1,
label='Density')
ax1.set_ylim([ymin,ymax])
plt.margins(0.05, 0.1)
plt.title('Density and Salinity')
fig.autofmt_xdate(rotation=50)
```

Define second y-axis

```
ax2 = plt.twinx()
plt.ylabel(u'Salinity')
plt.tick_params(axis='y')
s2,=ax2.plot(dtime, salin, 'r-', linewidth=1,
label='Salinity')
ax2.set_ylim([33,42])
```

Define the plot legend

```
lns = [s1, s2]
ax1.legend(handles=lns, loc='upper right')
```

Save and show the plot

```
plt.grid()
plt.tight_layout()
plt.savefig('/path-to-figure-folder/name.png', format='png',
dpi=300, transparent=False)
plt.show()
```

5.10 Barplots in Python

A barplot for a single dependent variable versus a factor variable (Fig. 5.23) [7]:
Load required libraries

```
import matplotlib.pyplot as plt
import numpy as np
import pandas as pd
import os
os.chdir('/path-to-local-data-folder/')
```

Load data

```
df = pd.read_csv('grouped_data.txt', delimiter='\t')
```

Define variables

```
mt = df['group'].unique()
lst=[]
err = []
for i in mt:
    i = np.mean(df.loc[df.group == i, :]['do'])
    lst.append(i)
for i in mt:
    i = np.std(df.loc[df.group == i, :]['do'])
    err.append(i)
```

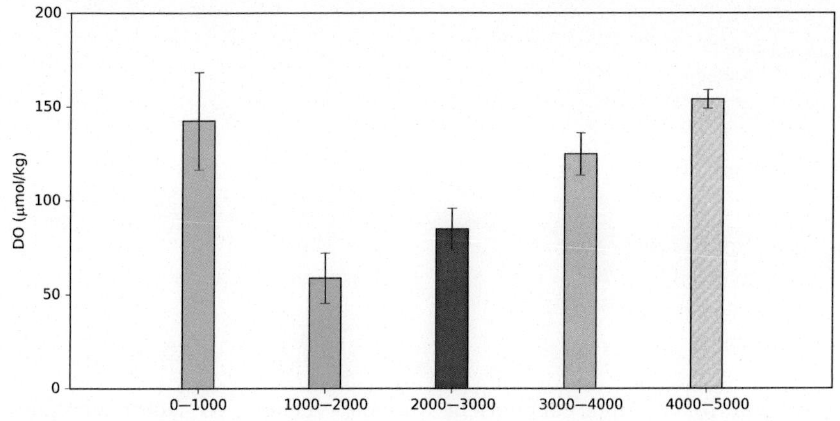

Fig. 5.23 A barplot for a single dependent variable versus a factor variable in Python.

Define plot parameters

```
ymin = 0
ymax = 200

labels = ['0-1000','1000-2000','2000-3000','3000-
4000','4000-5000']

colors = ['#97FFFF','#ADFF2F','#FF4500','#FFD700','#E0E0E0']

err_kw = dict(ecolor = 'black', capsize = 4)
```

Define positions of bars

```
pos=[]
width = 0.25
ind = range(1,len(mt)+1)

for i in ind:
    pos.append(i-width/2)
```

Define the plot

```
fig, ax = plt.subplots(figsize=(10,5))
plt.bar(pos, lst, width, yerr= err, error_kw=err_kw,
alpha=1, color=colors, label=mt)

ax.set_xticks(ind)
ax.set_xticklabels(labels)

ax.set_ylabel(r'DO ($\mu$mol/kg)')
ax.set_ylim([ymin,ymax])

ax.get_xaxis().tick_bottom()
ax.get_yaxis().tick_left()
ax.tick_params(direction='out')
```

Save and show the plot

```
plt.tight_layout()
plt.savefig('/path-to-figure-folder/name.png', format='png',
dpi=300, transparent=False)
plt.show()
```

A barplot for a single dependent variable versus two factor variables (Fig. 5.24):

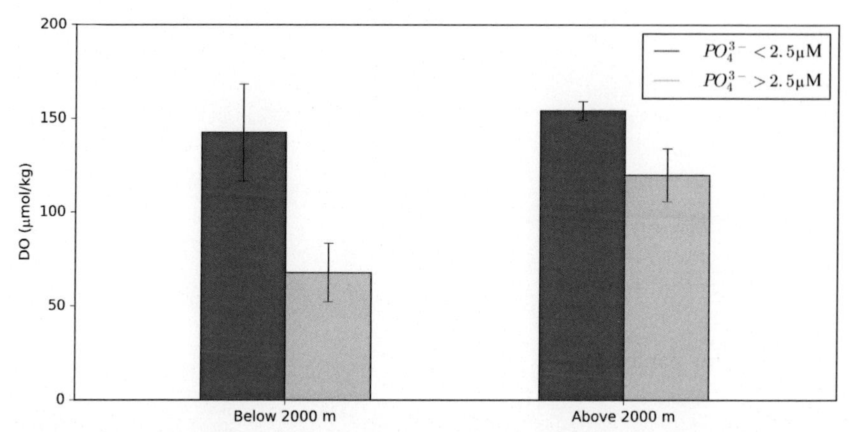

Fig. 5.24 A barplot for a single dependent variable versus two factor variables in Python.

Load required libraries

```
import matplotlib.pyplot as plt
import numpy as np
import pandas as pd
```

Load data

```
df = pd.read_csv('grouped_data.txt', delimiter='\t')
```

Define variables

```
m = np.mean(df.loc[df.phos25 == 'pb25'][df.limit ==
'bel2000']['do'])
n = np.mean(df.loc[df.phos25 == 'pa25'][df.limit ==
'bel2000']['do'])
p = np.mean(df.loc[df.phos25 == 'pb25'][df.limit ==
'ab2000']['do'])
r = np.mean(df.loc[df.phos25 == 'pa25'][df.limit ==
'ab2000']['do'])

alldata1 = [m,p]
alldata2 = [n,r]

err_m = np.std(df.loc[df.phos25 == 'pb25'][df.limit ==
'bel2000']['do'])
```

Continued

```
err_n = np.std(df.loc[df.phos25 == 'pa25'][df.limit ==
'bel2000']['do'])
err_p = np.std(df.loc[df.phos25 == 'pb25'][df.limit ==
'ab2000']['do'])
err_r = np.std(df.loc[df.phos25 == 'pa25'][df.limit ==
'ab2000']['do'])

err1 = [err_m,err_p]
err2 = [err_n,err_r]
```

Define plot parameters

```
ymin = 0
ymax = 200

labels = ['Below 2000 m','Above 2000 m']

color1 = ['#FF4500','#FF4500']
color2 = ['#FFD700','#FFD700']

err_kw = dict(ecolor = 'black', capsize = 4)
```

Define positions of bars

```
mt = df['limit'].unique()
pos1 = []
pos2 = []
width = 0.25
ind = range(1,len(mt)+1)
xmin = ind[0]-(2*width)
xmax = ind[1]+(3*width)

for i in ind:
    pos1.append(i-width/2)
    pos2.append(i+width/2)
```

Define the plot

```
fig, ax = plt.subplots(figsize=(10,5))

plt.bar(pos1, alldata1, width, yerr=err1, error_kw=err_kw,
alpha=1, color=color1)
plt.bar(pos2, alldata2, width, yerr=err2, error_kw=err_kw,
alpha=1, color=color2)

ax.set_xticks(pos2)
ax.set_xticklabels(labels)
```

```
ax.set_ylabel(r'DO ($\mu$mol/kg)')
ax.set_ylim([ymin,ymax])
ax.set_xlim([xmin,xmax])
ax.get_xaxis().tick_bottom()
ax.get_yaxis().tick_left()
ax.tick_params(direction='out')
```

Add legend

```
plt.plot([], c='#FF4500', label=r'PO4 < 2.5 uM')
plt.plot([], c='#FFD700', label=r'PO4 > 2.5 uM')
plt.legend()
```

Save and show the plot

```
plt.tight_layout()
plt.savefig('/path-to-figure-folder/name.png', format='png',
dpi=300, transparent=False)
plt.show()
```

A barplot for a single dependent variable (including positive and negative values) versus a factor variable (Fig. 5.25):

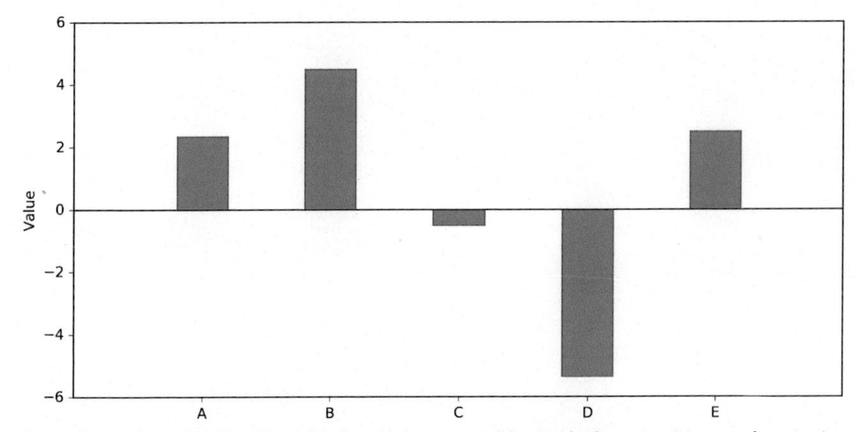

Fig. 5.25 A barplot for a single dependent variable (including positive and negative values) versus a factor variable in Python.

Load required libraries

```
import matplotlib.pyplot as plt
import numpy as np
import pandas as pd
import os
os.chdir('/path-to-local-data-folder/')
```

Load data

```
df = pd.read_csv('grouped_data.txt', delimiter='\t')
```

Define variables

```
mt = df['group'].unique()

lst=[]

for i in mt:
    i = np.mean(df.loc[df.group == i, :]['calc'])
    lst.append(i)
```

Define plot parameters

```
ymin = -6
ymax = 6
```

Define positions of bars

```
pos=[]
width = 0.4
ind = range(1,len(mt)+1)

for i in ind:
    pos.append(i-width/2)
```

Define the plot

```
fig, ax = plt.subplots(figsize=(10,5))

plt.bar(pos, lst, width, alpha=0.5, color='#00008B',
label=mt)
ax.axhline(y=0, xmin=0, xmax=1, color='black')

ax.set_xticks(ind)
ax.set_xticklabels(mt)
```

```
ax.set_ylabel('Value')
ax.set_ylim([ymin,ymax])

ax.get_xaxis().tick_bottom()
ax.get_yaxis().tick_left()
ax.tick_params(direction='out')
```

Save and show the plot

```
plt.tight_layout()
plt.savefig('/path-to-figure-folder/name.png', format='png',
dpi=300, transparent=False)
plt.show()
```

5.11 Boxplots in Python

A boxplot for a single dependent variable versus a factor variable (Fig. 5.26) [7]:

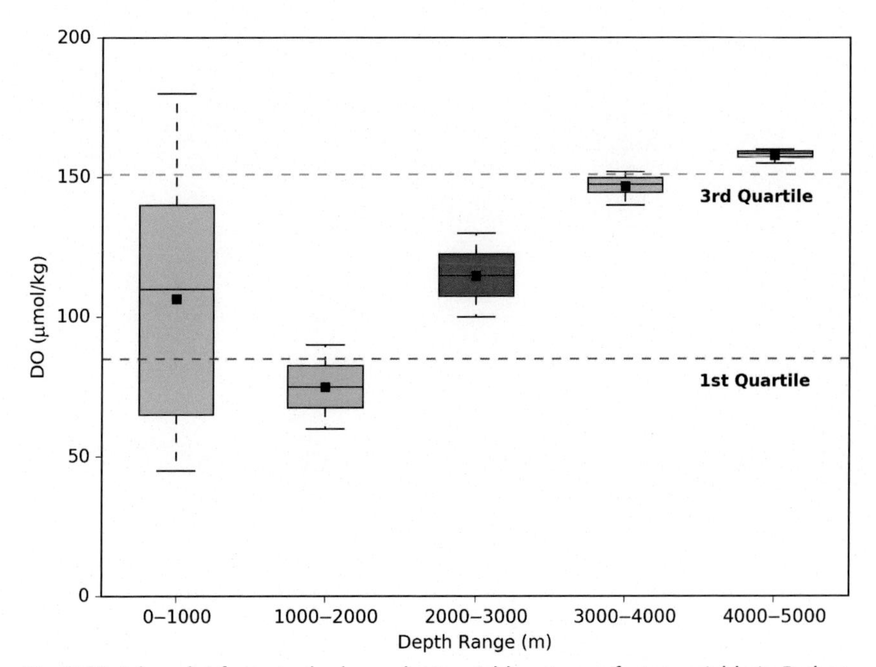

Fig. 5.26 A boxplot for a single dependent variable versus a factor variable in Python.

Load required libraries

```
import numpy as np
import pandas as pd
import matplotlib.pyplot as plt
import os
os.chdir('/path-to-local-data-folder/')
```

Load data

```
df = pd.read_csv('grouped_data.txt', delimiter='\t')
```

Define variables

```
mt = df['range'].unique()

lst1 = []
lst2 = []
lst3 = []
lst4 = []
lst5 = []
n = []

for i in mt:
    if i == 1000:
        n = df.loc[df.range == 1000, :]['do']
        lst1.append(n)
    elif i == 2000:
        n = df.loc[df.range == 2000, :]['do']
        lst2.append(n)
    elif i == 3000:
        n = df.loc[df.range == 3000, :]['do']
        lst3.append(n)
    elif i == 4000:
        n = df.loc[df.range == 4000, :]['do']
        lst4.append(n)
    else:
        n = df.loc[df.range == 5000, :]['do']
        lst5.append(n)

alldata = [lst1,lst2,lst3,lst4,lst5]
```

Define plot parameters

```
ymin = 0
ymax = 200

y1, y2 = np.percentile(df['do'], [25,75])

colors = ['#97FFFF','#ADFF2F','#FF4500','#FFD700','#E0E0E0']
labels = ['0-1000','1000-2000','2000-3000','3000-
4000','4000-5000']
meancol = dict(markerfacecolor='black')
mediancol = dict(color='black')
whiskercol = dict(color='black')
```

Define the plot

```
fig1 = plt.figure()
ax1 = fig1.add_subplot(111)
bplot = ax1.boxplot(alldata,vert=True,patch_artist=True,
showmeans=True,meanprops=meancol,medianprops=mediancol,
whiskerprops=whiskercol,labels=labels)
ax1.axhline(y=y1, xmin=0, xmax=1, color='red', linestyle='--')
ax1.axhline(y=y2, xmin=0, xmax=1, color='green', linestyle='--')
plt.text(4.5,y1-10,'1st Quartile',fontsize=11,
fontweight='bold')
plt.text(4.5,y2-10,'3rd Quartile',fontsize=11,
fontweight='bold')

ax1.set_ylabel(r'DO ($\mu$mol/kg)')
ax1.set_ylim([ymin,ymax])
ax1.set_xlabel('Depth Range (m)')

ax1.set_xticklabels(labels)

ax1.get_xaxis().tick_bottom()
ax1.get_yaxis().tick_left()
ax1.tick_params(direction='out')
```

Set colors for boxes

```
for patch, color in zip(bplot['boxes'], colors):
    patch.set_facecolor(color)
    patch.set_edgecolor('black')
```

Save and show the plot

```
plt.tight_layout()
plt.savefig('/path-to-figure-folder/name.png', format='png',
dpi=300, transparent=False)
plt.show()
```

A boxplot for a single dependent variable versus two factor variables (Fig. 5.27):

Load required libraries

```
import matplotlib.pyplot as plt
import numpy as np
import pandas as pd
import os
os.chdir('/path-to-local-data-folder/')
```

Load data

```
df = pd.read_csv('grouped_data.txt', delimiter='\t')
```

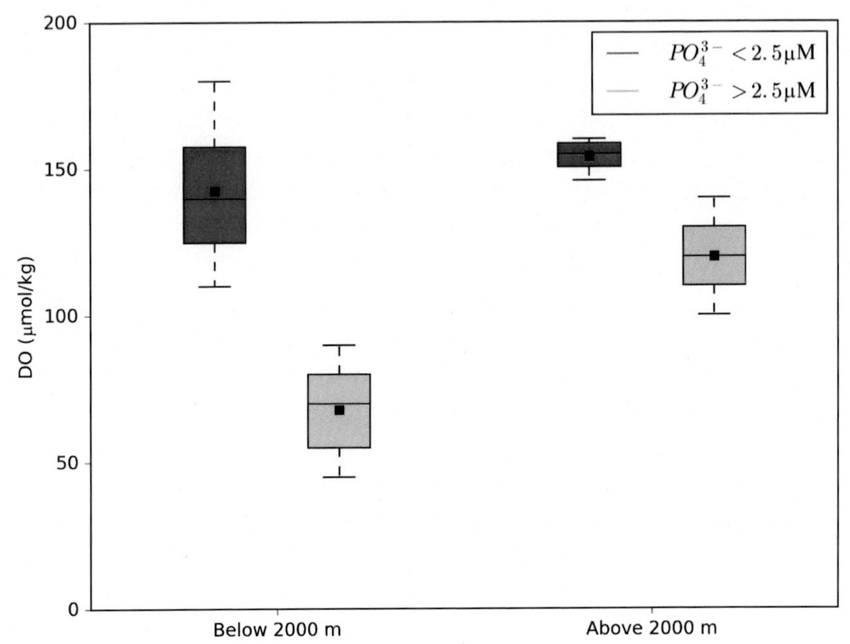

Fig. 5.27 A boxplot for a single dependent variable versus two factor variables in Python.

Define variables

```
m = df.loc[df.phos25 == 'pb25'][df.limit == 'bel2000']['do']
n = df.loc[df.phos25 == 'pa25'][df.limit == 'bel2000']['do']
p = df.loc[df.phos25 == 'pb25'][df.limit == 'ab2000']['do']
r = df.loc[df.phos25 == 'pa25'][df.limit == 'ab2000']['do']

alldata1 = [m,p]
alldata2 = [n,r]
```

Define plot parameters

```
labels = ['Below 2000 m','Above 2000 m']

ymin = 0
ymax = 200

color1 = ['#FF4500','#FF4500']
color2 = ['#FFD700','#FFD700']
meancol = dict(markerfacecolor='black')
mediancol = dict(color='black')
whiskercol = dict(color='black')
```

Define the plot

```
fig1 = plt.figure()
ax1 = fig1.add_subplot(111)
bplot1 = ax1.boxplot(alldata1,vert=True,patch_artist=True,
showmeans=True,meanprops=meancol,medianprops=mediancol,
whiskerprops=whiskercol,labels=labels,positions=[1,4],sym='',
widths=0.5)
bplot2 = ax1.boxplot(alldata2,vert=True,patch_artist=True,
showmeans=True,meanprops=meancol,medianprops=mediancol,
whiskerprops=whiskercol,labels=labels,positions=[2,5],sym='',
widths=0.5)

ax1.set_ylabel(r'DO ($\mu$mol/kg)')
ax1.set_ylim([ymin,ymax])
ax1.set_xlim([0,6])
ax1.set_xticks([1.5, 4.5])

ax1.set_xticklabels(labels)

ax1.get_xaxis().tick_bottom()
ax1.get_yaxis().tick_left()
ax1.tick_params(direction='out')
```

Add legend

```
plt.plot([], c='#FF4500', label=r'PO4 < 2.5 uM')
plt.plot([], c='#FFD700', label=r'PO4 > 2.5 uM')
plt.legend()
```

Set colors for boxes

```
for patch, color in zip(bplot1['boxes'], color1):
    patch.set_facecolor(color)
    patch.set_edgecolor('black')
for patch, color in zip(bplot2['boxes'], color2):
    patch.set_facecolor(color)
    patch.set_edgecolor('black')
```

Save and show the plot

```
plt.tight_layout()
plt.savefig('/path-to-figure-folder/name.png', format='png',
dpi=300, transparent=False)
plt.show()
```

5.12 Pie charts in Python

A 2d pie chart (Fig. 5.28) [7]:

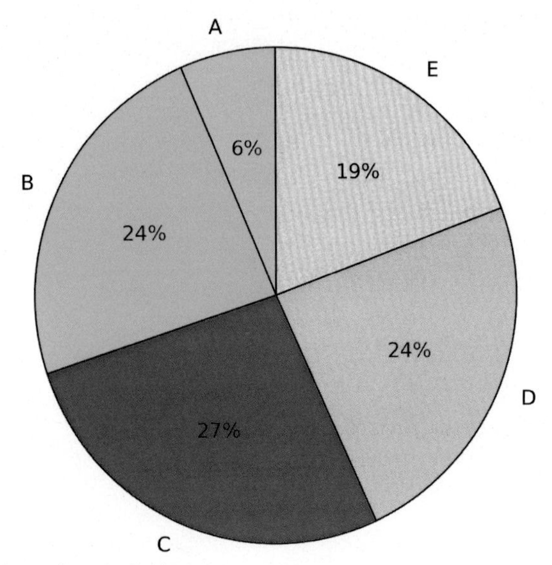

Fig. 5.28 A 2d pie chart in Python.

Load required libraries

```
import matplotlib.pyplot as plt
import numpy as np
import pandas as pd
import os
os.chdir('/path-to-local-data-folder/')
```

Load data

```
df = pd.read_csv('grouped_data.txt', delimiter='\t')
```

Define variables

```
mt = df['group'].unique()
lst=[]
for i in mt:
    i = np.mean(df.loc[df.group == i, :]['si'])
    lst.append(i)
explode = (0, 0, 0, 0, 0)
colors = ['#ADFF2F','#97FFFF','#FF4500','#FFD700','#E0E0E0']
```

Define the plot

```
fig, ax = plt.subplots(figsize=(10,5))
ax.pie(lst,  explode=explode,  labels=mt,  autopct='%1.0f%%',
shadow=False, startangle=90, colors=colors)
ax.axis('equal')
```

Save and show the plot

```
plt.savefig('/path-to-figure-folder/name.png', format='png',
dpi=300, transparent=False)
plt.show()
```

5.13 3D plots in Python

A 3d scatter plot (Fig. 5.29) [7]:

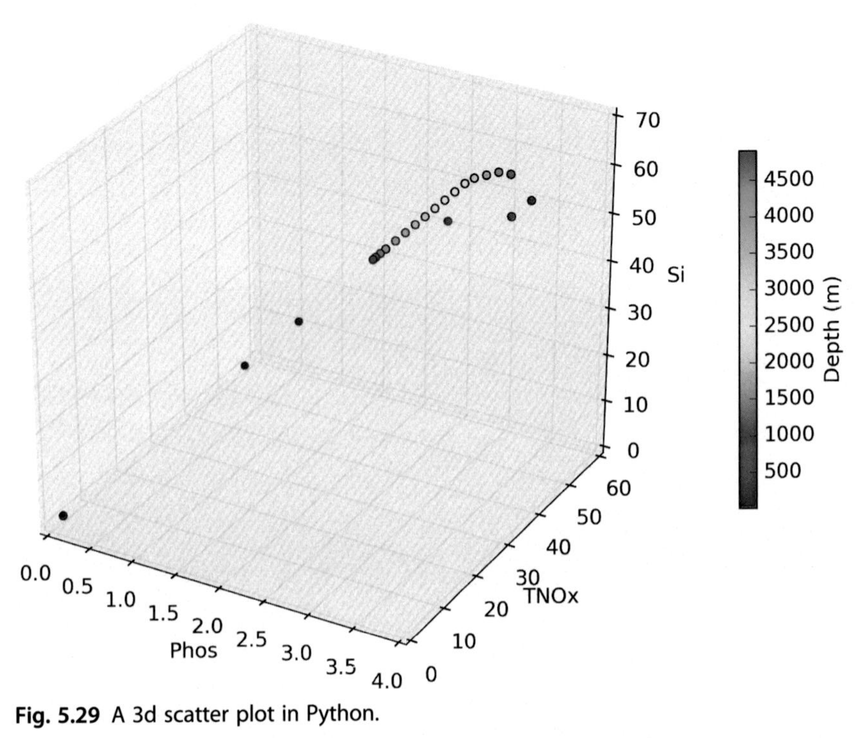

Fig. 5.29 A 3d scatter plot in Python.

Load required libraries

```
from mpl_toolkits.mplot3d import Axes3D
import matplotlib.pyplot as plt
import numpy as np
import matplotlib.cm as cm
```

Load data

```
f = open('/path-to-local-data-folder/data.txt', 'r')
data = np.genfromtxt(f, dtype="float", delimiter='\t',
names=True)
f.close()
```

Define variables

```
depth = data['depth']
phos = data['phos']
```

```
tnox = data['tnox']
si = data['si']
del(data)
```

Define the plot

```
fig = plt.figure()
ax = fig.add_subplot(111, projection='3d')

scat = ax.scatter(phos, tnox, si, c=depth, marker='o',
cmap=cm.Spectral)

cbar = fig.colorbar(scat, shrink=0.5, aspect=20)
cbar.ax.set_ylabel('Depth (m)')

ax.set_xlabel('Phos')
ax.set_ylabel('TNOx')
ax.set_zlabel('Si')

ax.set_xlim([0,4])
ax.set_ylim([0,60])
ax.set_zlim([0,70])
```

Save and show the plot

```
plt.tight_layout()
plt.savefig('/path-to-figure-folder/name.png', format='png',
dpi=300, transparent=False)
plt.show()
```

A 3d surface plot (Fig. 5.30):
Load required libraries

```
from mpl_toolkits.mplot3d import Axes3D
import matplotlib.pyplot as plt
import numpy as np
import matplotlib.cm as cm
```

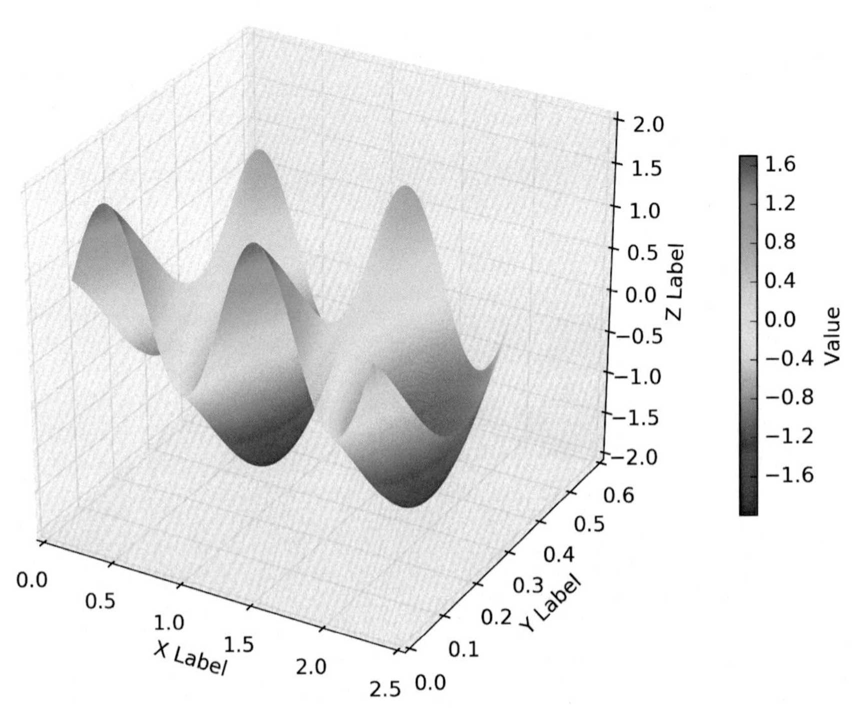

Fig. 5.30 A 3d surface plot in Python.

Define variables

```
n = 10000
pi = np.pi

X = np.arange(0, 2*pi, 2*pi/n)
Y = np.arange(pi/12, pi/2, 2*pi/n)
M, N = np.meshgrid(X, Y)
R = np.sin(2*M)+np.cos(3*N)
T = M/3
S = N/3
Z = R
```

Define the plot

```
fig = plt.figure()
ax = fig.add_subplot(111, projection='3d')

surf = ax.plot_surface(T, S, Z, cmap=cm.Spectral, linewidth=0,
antialiased=True)

cbar = fig.colorbar(surf, shrink=0.5, aspect=20)
cbar.ax.set_ylabel('Value')
```

```
ax.set_xlabel('X Label')
ax.set_ylabel('Y Label')
ax.set_zlabel('Z Label')
```

Save and show the plot

```
plt.tight_layout()
plt.savefig('/path-to-figure-folder/name.png', format='png',
dpi=300, transparent=False)
plt.show()
```

5.14 Ternary plots in Python

Ternary plots can be plotted by using 'ternary' library. The data should be transformed before use in this plot. Sums of three variables should be 100. For this purpose, Ternplot [8] excel spreadsheet or similar ternary plotting tools can be used.
A ternary plot (Fig. 5.31):

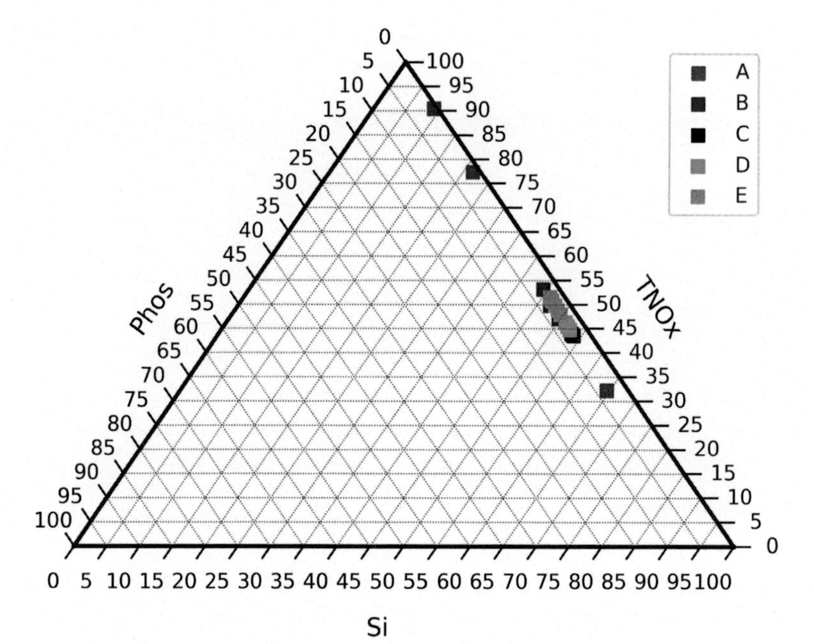

Fig. 5.31 A ternary plot in Python.

Load required libraries

```
import ternary
import random
import numpy as np
from matplotlib import pyplot as plt
```

Load data

```
points = []
with open('/path-to-local-data-folder/tern_data.txt') as
handle:
    for line in handle:
        points.append(list(map(float, line.split('\t'))))
```

Define the plot

```
scale = 100
figure, tax = ternary.figure(scale=scale)
```

Define plot boundary and grids

```
tax.boundary(linewidth=2.0)
tax.gridlines(color="blue", multiple=5)
```

Define axis labels

```
fontsize = 12
offset = 0.14
tax.set_title('', fontsize=fontsize)
tax.left_axis_label('Phos', fontsize=fontsize, offset=offset)
tax.right_axis_label('TNOx', fontsize=fontsize, offset=offset)
tax.bottom_axis_label('Si', fontsize=fontsize, offset=offset)
```

Plot scatter points

```
tax.scatter(points[0:3], marker='s', color='red', label='A')
tax.scatter(points[4:7], marker='s', color='blue',
label='B')
```

```
tax.scatter(points[8:11], marker='s', color='black',
label='C')
tax.scatter(points[12:15], marker='s', color='green',
label='D')
tax.scatter(points[16:22], marker='s', color='gray',
label='E')

tax.ticks(axis='lbr', multiple=5, linewidth=1, offset=0.025)
tax.legend()
tax.get_axes().axis('off')
tax.clear_matplotlib_ticks()
```

Save and show the plot

```
tax.savefig('/path-to-figure-folder/name.png', format='png',
dpi=600)
tax.show()
```

References

[1] R Help Documentation, Can be accessed from R console by typing "help.start()".
[2] A. Kassambara, ggpubr: 'ggplot2' Based Publication Ready Plots, R package version 0.2, https://cran.r-project.org/package=ggpubr, 2018.
[3] J. Lemon, Plotrix: a package in the red light district of R, R-News 6 (4) (2006) 8–12.
[4] K. Soetaert, OceanView: Visualisation of Oceanographic Data and Model Output, R package version 1.0.4, https://cran.r-project.org/package=OceanView, 2016.
[5] K. Soetaert, plot3D: Plotting Multi-Dimensional Data, R package version 1.1.1, https://cran.r-project.org/package=plot3D, 2017.
[6] N. Hamilton, ggtern: An Extension to 'ggplot2', for the Creation of Ternary Diagrams, R package version 3.1.0, https://cran.r-project.org/package=ggtern, 2018.
[7] Python Documentation, Matplotlib Gallery of Example Plots, https://matplotlib.org/gallery/index.html. Accessed 16 January 2019.
[8] D. Marshall, Ternplot: an Excel spreadsheet for ternary diagrams, Comput. Geosci. 22 (6) (1996) 697–699.

Index

Note: Page numbers followed by *f* indicate figures, *t* indicate tables, and *b* indicate boxes.

Printed in the United States
By Bookmasters